The Bethesda System for Reporting Cervical Cytology

Second Edition

The Bethesda System for Reporting Cervical Cytology

Definitions, Criteria, and Explanatory Notes

Second Edition

Editors

Diane Solomon, MD
Senior Investigator
Breast and Gynecologic
   Cancer Research Group
Division of Cancer Prevention
National Cancer Institute
NIH, DHHS
Bethesda, Maryland

Ritu Nayar, MD
Associate Professor
Director of Cytopathology
Department of Pathology
Northwestern University
Feinberg School of Medicine
Chicago, Illinois

With a Foreword by Robert J. Kurman, MD

With an Introduction by Diane D. Davey, MD
and David C. Wilbur, MD

 Springer

Diane Solomon, MD
National Cancer Institute
6130 Executive Blvd.
Rockville, MD 20852
USA

Ritu Nayar, MD
Northwestern University
Feinberg School of Medicine
Department of Pathology, Ward 6-204
303 E. Chicago, IL 60611
USA

Cover Illustrations (From left to right):

Front, upper panel: Endocervical cells, liquid based preparation; Atypical squamous cells of undetermined significance (ASC-US), conventional preparation; Endocervical adenocarcinoma, conventional preparation.

Front, lower panel: Low grade squamous intraepithelial lesion (LSIL), liquid based preparation; Tubal metaplasia, conventional preparation; Adenocarcinoma, extrauterine (metastatic gastric carcinoma), conventional preparation.

Back: Squamous cell carcinoma, conventional preparation; High grade squamous intraepithelial lesion (HSIL), conventional preparation; Low grade squamous intraepithelial lesion (LSIL), anal-rectal sample, liquid based preparation.

Library of Congress Cataloging-in-Publication Data
The Bethesda system for reporting cervical cytology / [edited by] Diane Solomon, Ritu Nayar.
   p. ; cm.
  Includes bibliographical references and index.
  ISBN 0-387-40358-2 (s/c : alk. paper)
  1. Uterus—Cancer. 2. cytodiagnosis. I. Solomon, Diane, 1955– II. Nayar, Ritu.
  [DNLM: 1. Cervical Intraepithelial Neoplasia—classification. 2. Cervical Intraepithelial
Neoplasia—pathology. 3. Cervix Dysplasia—diagnosis. 4. Cervix Dysplasia—pathology. 5.
Cervix Neoplasms—diagnosis. 6. Cervix Neoplasms—pathology. 7.
Cytodiagnosis—standards. WP 480 B562 2003]
  RC280.U8B48 2003
  616.99′466—dc 22   2003058441

ISBN 978-0-387-40358-8
ISBN 0-387-40358-2   Printed on acid-free paper.

Printed in China.

9 8 7 6 5

springer.com

Foreword

The Bethesda System: A Historical Perspective

In December 1988, a small group of individuals with expertise in cytology, histopathology, and patient management participated in an NCI-sponsored meeting in Bethesda, Maryland, with the goal of developing a system for reporting Pap smears that would communicate the cytology interpretation to the clinician in a clear and relevant fashion. Before that time, laboratories generally utilized a numeric "Pap Class" system of reporting results, which was confusing and often idiosyncratic, or the "dysplasia" terminology that had poor interobserver reproducibility in practice.

The result of this first meeting was The 1988 Bethesda System (TBS). This new terminology reflected three fundamental principles:

1. The terminology must communicate clinically relevant information from the laboratory to the patient's health-care provider.
2. The terminology should be uniform and reasonably reproducible across different pathologists and laboratories, and also flexible enough to be adapted in a wide variety of laboratory settings and geographic locations.
3. The terminology must reflect the most current understanding of cervical neoplasia.

The new nomenclature was initially met with skepticism by many, not only because it proposed replacing classifications that had been in place for several decades but also because it eliminated the diagnostic category of moderate dysplasia or CIN 2. Traditionally, cellular changes of human papillomavirus (HPV) ("koilocytotic atypia") were considered separate from "true" cervical cancer precursors, which were subdivided into four (mild, moderate, severe dysplasia, and carcinoma in situ) or three (CIN 1, 2, 3) categories, reflecting what was perceived to be a biologic continuum. TBS proposed a bipartite division, low-grade and high-grade squamous intraepithelial lesions (LSIL and HSIL) (See Fig. F.1 on page vii). The rationale for reducing (or consolidating) multiple categories of HPV effect, degrees of dysplasia, or grades of CIN to the two tiers of LSIL and HSIL was based on the principles of the Bethesda System just stated:

a. LSIL/HSIL reflected clinical decision thresholds at the time; LSIL was often followed, but HSIL triggered colposcopic evaluation.

b. The reduced number of diagnostic categories improved interobserver variability and intraobserver reproducibility.

c. Research suggested that the biology of cervical abnormalities might not be as linear and continuous as the spectrum of morphologic changes would imply.

Of all the changes introduced by TBS, probably none was as problematic and controversial as the infamous "atypical squamous cells of undetermined significance," or "ASCUS." ASCUS highlights the inherent tension that exists among pathologists who are not always able to make black-and-white decisions on cellular or histologic specimens and clinicians whose management decisions are more clearly dichotomous—to treat or not to treat. ASCUS, a true reflection of the cytopathologist's inability to make a definitive diagnosis in certain cases, often left clinicians feeling compelled to evaluate the patient by colposcopy, a time-consuming and expensive procedure. Because ASCUS was reported in approximately 2.5 million Pap tests in the United States annually, this was a major problem. Furthermore, because the management of women with ASCUS results was unclear, the NCI sponsored a clinical trial, the ASCUS/LSIL Triage Study (ALTS), to determine the best management for these patients.

The results of ALTS provided data for the development of evidence-based guidelines for management of women with abnormal cytology results (using Bethesda terminology) under the auspices of the American Society for Colposcopy and Cervical Pathology (ASCCP). Among the major findings of ALTS and other studies was that the spectrum of morphologic changes that was believed to constitute the pre-invasive phase of cervical cancer consists of two biologically different conditions: a viral infection caused by HPV that results in a low-grade squamous intraepithelial lesion and an HPV-induced cervical cancer precursor, a high-grade squamous intraepithelial lesion. These findings confirmed the validity of the bipartite LSIL/HSIL classification in TBS.

ALTS, initiated in response to issues raised by TBS, has also set into motion a new approach to cervical cancer screening that utilizes molecular testing for HPV to triage women with ASCUS, as it is more sensitive and very likely more cost-effective for the detection of underlying high-grade lesions than repeat cytology. In the future, cervical cancer screening may well begin with HPV testing with cytology as a triage for HPV-positive samples. This change in screening will obviously have an enormous impact on cytology. On the one hand, the absolute numbers of cervical cytology specimens would decrease, but on the other hand cytologic evaluation, in its new role as a method of triage, would become even more important and challenging.

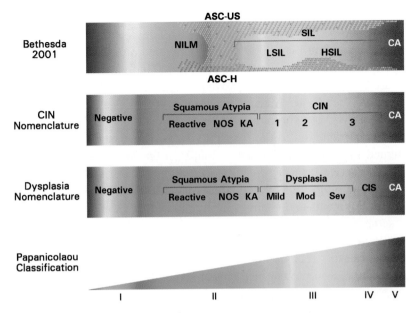

Comparison of four cytologic classifications for squamous cells: Bethesda 2001, CIN Nomenclature, Dysplasia Nomenclature, and Papanicolaou Classification (the categories not shown to scale). See the end of the legend for the explanation of abbreviations. The regions are represented by **graded shades** of colors to emphasize the morphologic continuum of cytologic findings and the indiscrete transitions between categories; shades of **blue** = negative; **green** = equivocal; **yellow** = low-grade intraepithelial abnormalities (roughly HPV infection); **orange** = high-grade intraepithelial abnormalities (roughly HPV-related intraepithelial neoplasia); and **red** = carcinoma. The Papanicolaou Classification was based on risk of a patient harboring cancer and is represented diagramatically as a wedge of increasing risk from class I to class V.

2001 Bethesda System: The categories of ASC-US and ASC-H are represented by **dots** that span multiple categories, emphasizing the nonhierarchical nature of these categories. Most ASC-US reflects difficulties in the distinction between reactive changes and LSIL, whereas most ASC-H reflects the differential between reactive (immature) metaplasia and HSIL. SIL remains dichotomized; LSIL implies changes that mainly reflect HPV infection (eliminating the distinction between KA and CIN1), whereas HSIL implies a higher risk lesion, including cancer precursors.

Abbreviations used: CIN = cervical intraepithelial neoplasia; ASC-US = atypical squamous cells of undetermined significance; ASC-H = atypical squamous cells, cannot exclude an HSIL; NILM = negative for intraepithelial lesion and malignancy; ASC = atypical squamous cells; SIL = squamous intraepithelial lesion; LSIL = low-grade squamous intraepithelial lesion; HSIL = high-grade squamous intraepithelial lesion; KA = koilocytotic atypia (HPV effect); HPV = human papillomavirus; CA = invasive carcinoma; NOS = not otherwise specified; Mild = mild dysplasia; Mod = moderate dysplasia; Sev = severe dysplasia; and CIS = carcinoma *in situ*.

The most recent Bethesda Workshop, held in 2001, utilized the Internet to widen participation in the process of reviewing and revising the terminology. More than 2000 comments were considered before the actual meeting, which brought together over 400 individuals from more than two dozen countries. The innovative use of the Internet continues with an interactive Web-based Atlas of images that serves as a companion to this book at www.cytopathology.org/NIH.

In the introduction to the first TBS Atlas, it was stated that TBS was designed to be flexible so that it could evolve in response to changing needs in cervical cancer screening as well as to advances in the field of cervical pathology. It is paradoxical that instead of TBS responding to new developments and changes in the field of cervical carcinogenesis, it has actually led the way in many areas. Thus, TBS has played a vital role in initiating research in the biology of cervical cancer, in exploring new approaches and strategies in patient management, and in incorporating new technologies into cervical cancer screening.

Robert J. Kurman, MD
Baltimore, Maryland
August 2003

Acknowledgments

I. For Image Contribution

George G. Birdsong: 1.1, 1.2, 1.3, 1.4, 1.5, 1.9, 1.18

Charlotte Brahm (courtesy of Cytyc Corporation): 6.28

Sally-Beth Buckner: 2.10, 2.30

Terence J. Colgan: 2.12 (insert), 2.39

Jamie L. Covell: 1.11, 1.17; 2.2, 2.9, 2.14, 2.16, 2.22, 2.28, 2.36 (right), 2.37 (left); 3.2, 3.7; 4.17; 5.14, 5.21, 5.27, 5.31, 5.32; 6.1, 6.3, 6.7, 6.9 (left), 6.10, 6.11, 6.22, 6.25, 6.27, 6.29, 6.32, 6.33, 6.36, 6.37, 6.40, 6.42; 7.1, 7.2, 7.3, 7.4, 7.5, 7.6, 7.7, 7.8, 7.9

Teresa M. Darragh: 8.1, 8.2, 8.3, 8.4, 8.5, 8.6, 8.7

Diane D. Davey: 2.12; 5.4, 5.19, 5.37 (left)

Denise V.S. DeFrias: 5.25

Rose Marie Gatscha: 5.41

Michael Henry: 1.6, 1.7, 1.8, 1.10; 2.3 (right)

Ronald D. Luff: 5.36

Ann T. Moriarty: 2.35; 3.3, 3.5, 3.8 (insert)

Ritu Nayar: 1.15; 2.8, 2.9 (insert), 2.21, 2.31, 2.32, 2.37 (right), 2.38, 2.40; 3.8; 4.4, 4.16, 4.20; 5.2, 5.3, 5.6, 5.7, 5.9, 5.20, 5.29, 5.35, 5.38; 6.30, 6.35

Celeste N. Powers: 4.9

Mark E. Sherman: F.1, 2.13, 2.27; 4.1, 4.5, 4.8, 4.11, 4.13, 4.14; 5.33; 6.6, 6.18, 6.24, 6.39; 7.2 (insert), 7.8 (insert)

Mary K. Sidaway: 4.12

Diane Solomon: 1.16; 2.1, 2.4, 2.7, 2.11, 2.15, 2.19, 2.20, 2.23, 2.24, 2.25, 2.29, 2.33, 2.34; 3.1, 3.6; 4.2, 4.3, 4.7, 4.10, 4.18, 4.19; 5.1, 5.8, 5.10, 5.12, 5.15, 5.16, 5.17, 5.18, 5.22, 5.23, 5.24, 5.34, 5.40; 6.8, 6.9 (right), 6.14, 6.16, 6.17, 6.19, 6.20, 6.21, 6.26, 6.31, 6.34, 6.41

Sana O. Tabbara: 4.21

David C. Wilbur: 1.12, 1.13, 1.14; 2.3 (left), 2.5, 2.17; 3.4; 4.6, 4.15; 5.5, 5.11, 5.13, 5.26, 5.28, 5.30, 5.37 (right), 5.39; 6.2, 6.4, 6.5, 6.12, 6.13, 6.15, 6.23, 6.38

Nancy A. Young: 2.6, 2.18, 2.23 (insert), 2.26, 2.36 (left)

II. Acknowledgment for Image Selection

ASC-NCI Bethesda Task Force:

ASC: Ritu Nayar, MD (Chair), George G. Birdsong, MD, Jamie L. Covell, BSc, CT (ASCP), Ann T. Moriarty, MD, Dennis M. O'Connor, MD, Marianne U. Prey, MD, Stephen S. Raab, MD, Mark E. Sherman, MD, Sana O. Tabbara, MD, Thomas C. Wright, MD, Nancy A. Young, MD

NCI: Diane Solomon, MD (Chair)

Consultants: David C. Wilbur, MD and Diane D. Davey, MD

Web site for image selection:

Michael Montgomery, Terrapin Systems, Bethesda, MD

Brandon K. Winbush: Information Technology consultant for Department of Pathology, Northwestern University, Chicago, IL

Statistics:

Stephen S. Raab, MD

III. Acknowledgment for Manuscript Review

The editors would also like to thank Drs. Diane D. Davey and David C. Wilbur for their review of the manuscript and their invaluable feedback.

Contents

Contributors

Fadi W. Abdul-Karim, MD, Department of Pathology, University Hospitals of Cleveland and Case Western Reserve University, Cleveland, OH 44106, USA

Jonathan S. Berek, MD, MMSc, Department of Obstetrics & Gynecology, UCLA Women's Reproductive Cancer Program, David Geffen School of Medicine at UCLA, Los Angeles, CA 90095-1740, USA

Marluce Bibbo, MD, Department of Pathology, Jefferson Medical College, Philadelphia, PA 19107, USA

George G. Birdsong, MD, Department of Pathology and Laboratory Medicine, Emory University School of Medicine and Department of Anatomic Pathology, Grady Health System, Atlanta, GA 30322 USA

Sally-Beth Buckner, SCT(ASCP), Department of Cellular Pathology and Genetics, Armed Forces Institute of Pathology, Washington, D.C. 20306, USA

David C. Chhieng MD, Department of Pathology, University of Alabama at Birmingham, Birmingham, AL 35249-6823, USA

Edmund S. Cibas, MD, Department of Pathology, Brigham and Women's Hospital, Harvard Medical School, Boston, MA 02115, USA

Terence J. Colgan, MD, Department of Laboratory Medicine and Pathobiology, University of Toronto, Mount Sinai Hospital, Toronto, Ontario, M5G 1X5 Canada

Jamie L. Covell, BS CT (ASCP), Department of Pathology, University of Virginia Health System, Charlottesville, VA 22908, USA

Teresa M. Darragh, MD, University of California, San Francisco, Departments of Pathology and Ob/Gyn, San Francisco, CA 94143-1785, USA

Diane D. Davey, MD, Department of Pathology and Laboratory Medicine, University of Kentucky Medical Center, Lexington, KY 40536-0298, USA

Paul A. Elgert, CT(ASCP), CMIAC, Cytopathology Laboratory, Department of Pathology, Bellevue Hospital Center, New York University School of Medicine, New York, NY 10016, USA

Rose Marie Gatscha, SCT(ASCP), CFIAC, Cytology Service, Memorial Sloan-Kettering Cancer Center, New York, NY 10021, USA

Barbara Guidos, SCT (ASCP), The American Society for Clinical Pathology, Chicago, IL 60612, USA

Michael Henry, MD, Division of Laboratory Medicine, Cleveland Clinic Florida, Naples, FL 34119, USA

Kenneth R. Lee, MD, Department of Pathology Brigham & Women's Hospital, Harvard Medical School, Boston, MA 02115, USA

Ronald D. Luff, MD, MPH, Anatomic Pathology, Quest Diagnostics Incorporated, Teterboro, NJ 07608, USA

Dina R. Mody, MD, Department of Pathology, Baylor College of Medicine, Houston, TX 77030, USA

Ann T. Moriarty, MD, AmeriPath Indiana, Indianapolis, IN 46219, USA

Dennis M. O'Connor, MD, Clinical Associates, Louisville, KY 40207, USA

Celeste N. Powers, MD, PhD, Department of Pathology, Medical College of Virginia, Virginia Commonwealth University, Richmond, VA 23298-0139, USA

Marianne U. Prey, MD, Anatomic Pathology, Quest Diagnostics Incorporated, St. Louis, MO 63146, USA

Stephen S. Raab, MD, Department of Pathology, University of Pittsburgh, Pittsburgh, PA 15232, USA

Mark E. Sherman, MD, Hormonal and Reproductive Epidemiology Branch, Division of Epidemiology and Genetics, National Cancer Institute, Rockville, MD 20852, USA

Mary K. Sidawy, MD, Department of Pathology, The George Washington University, Washington, D.C. 20037, USA

Sana O. Tabbara, MD, Department of Pathology, The George Washington University, Washington, D.C. 20037, USA

David C. Wilbur, MD, Department of Pathology, Massachusetts General Hospital, Harvard Medical School, Boston, MA 02114, USA

Thomas C. Wright, MD, Division of Ob/Gyn Pathology, College of Physicians and Surgeons, Columbia University Presbyterian Medical Center, New York, NY 10031, USA

Nancy A. Young, MD, Department of Pathology, Fox Chase Cancer Center, Philadelphia, PA 19111, USA

List of Abbreviations

AIS:	adenocarcinoma in situ of endocervix
ALTS:	ASCUS/ LSIL Triage study
AGC:	atypical glandular cells
ASC:	atypical squamous cells
ASC:	American Society of Cytopathology
ASCCP:	American Society for Colposcopy and Cervical Pathology
ASC-H:	atypical squamous cells, cannot exclude high-grade squamous intraepithelial lesion
ASC-US:	atypical squamous cells of undetermined significance
BIRP:	Bethesda Interobserver Reproducibility Project
CIS:	carcinoma in situ
CIN:	cervical intraepithelial neoplasia
CP:	conventional preparation
EC/TZ:	endocervical/ transformation zone
EM's:	endometrial cells
HSIL:	high-grade squamous intraepithelial lesion
HPV:	human papillomavirus
IUD:	intrauterine device
LBP:	liquid-based preparation
LEEP:	loop electrosurgical excision procedure
LMP:	last menstrual period
LUS:	lower uterine segment
LSIL:	low-grade squamous intraepithelial lesion
MMMT:	malignant mixed mesodermal tumor
N:C or N/C:	nuclear to cytoplasmic ratio
NCI:	National Cancer Institute, Bethesda, MD
NILM:	negative for intraepithelial lesion or malignancy
SCUC:	small cell undifferentiated carcinoma
SIL:	squamous intraepithelial lesion
Spp:	species
TBS:	The Bethesda System
TCC:	transitional cell carcinoma
T zone:	transformation zone
VAIN:	vaginal intraepithelial neoplasia

Introduction

The second edition of the Bethesda System atlas represents the joint efforts of multiple individuals and organizations. Nearly 1 year before the 2001 Bethesda Workshop, nine forum groups, each consisting of 6 to 10 individuals, initiated a lengthy process designed to provide for the widest input possible via an Internet-based "discussion." Forty-four international organizations with interest in cervical cytopathology cosponsored the Bethesda System 2001 Workshop along with the National Cancer Institute (NCI). More than 400 individuals took part in the final working session, the third Bethesda workshop, in April 2001. As a follow-up to this conference and the publication of the revised 2001 Bethesda terminology,[1] conference moderator Dr. Diane Solomon approached the American Society of Cytopathology (ASC) regarding the development of a Bethesda Web site and an updated edition of the Bethesda "blue book" atlas. An ASC-NCI task force, named by Drs. Diane Davey and David Wilbur (at the time the President and President-Elect of the ASC), included the Bethesda forum moderators or their designees and selected consultants. Drs. Ritu Nayar and Diane Solomon jointly chaired this task force. The task force proposed working in parallel on the new edition of the atlas and an educational Web site featuring additional images beyond the number utilized in the atlas.

This new edition of the Bethesda System atlas expands on the popular features of the 1994 edition.[2] The atlas has been divided into chapters conforming to the major Bethesda interpretive categories. Each chapter consists of a background discussion, an in-depth description of cytologic criteria, and explanatory notes. New features include images of liquid-based preparations (LBPs), sample reports, references, and an index. Cytologic criteria are described in general for all specimen types in every chapter, with significant differences in LBPs noted following the general criteria when applicable. Note that Bethesda does not endorse any particular methodology for specimen collection.

Over 1,000 images were evaluated for this atlas. The majority went through a multistage review process. First, the relevant forum group evaluated images from the previous atlas and hundreds of new illustrations submitted by the ASC-NCI task force and Bethesda forum group members. Approximately 30% of images survived this initial review. Second, the images selected from the first round were placed on a Web site and scored independently by the ASC-NCI task force members. Only 58% of the images—those scoring above a selected mean threshold in the second round—were then included in the atlas. Third, before publication, approximately half the atlas images (indicated by an asterisk in the atlas legend) were posted as "unknowns" on a Web site open to the cytopathology community as part of the Bethesda Interobserver Reproducibility Project (BIRP). Hundreds of participants submitted their answers online, to provide a more realistic gauge of interpretive reproducibility. The resulting histograms of participants' interpretations of the images can be viewed at www.cytopathology.org/NIH.

The 186 illustrations in this atlas represent the spectrum of changes seen on both conventional smears and LBPs; 90% are new images and 40% are from LBPs. Some images represent classic examples of an entity whereas others, which were selected to illustrate interpretive dilemmas or "borderline" morphologic features, may not be interpreted in the same way by all cytologists. We hope that this revised atlas, which provides explanatory text and morphologic criteria, will help in laboratory implementation of the 2001 Bethesda System; however, some degree of interobserver and interlaboratory variability will always remain a reality.[3]

The 2001 Bethesda System includes changes that are based on clinical input and advances in the understanding of the biology of cervical cancer. The term "diagnosis" is replaced by "interpretation" or "result" in the heading of the cervical cytology report. Participants at the Bethesda 2001 conference agreed that cervical cytology should be viewed primarily as a "screening test, which in some instances may serve as a medical consultation by providing an interpretation that contributes to a diagnosis."[1] A patient's final diagnosis and management plan integrate not only the cervical cytology result but also history, clinical findings, and other laboratory results such as biopsy interpretations. This change in terminology emphasizes that the cytology result represents one component of, and may not always reflect, the final patient diagnosis.

The goal of the Bethesda System has always been to promote effective communication of relevant cytology findings between the laboratory and clinician to provide optimal patient care. New chapters discuss ancillary tests and computer-based interpretations that may be incorporated into the cytology report. The most common ancillary test used at present is human papillomavirus DNA testing, but any reporting framework should be applicable to other molecular tests that may be developed in the future.

The reporting of unsatisfactory specimens has also been clarified; the present terminology emphasizes that to consider a specimen "unsatisfactory," significant laboratory work must be performed to process and fully evaluate the specimen. The revised terminology for adequacy does allow for the mention of other pertinent findings (such as endometrial cells in a woman age 40 or older, or organisms) when the specimen is reported as being unsatisfactory.

The Bethesda System was developed primarily for cervical cytology specimens, and both the terminology and morphologic criteria reflect this. The term "cervical cytology" is used in this edition of the atlas (as opposed to "cervico-vaginal cytology") as the vagina is not specifically sampled with most cervical collection methods. However, specimens from other sites such as the vagina and anal-rectal samples may be reported using similar terminology. An entire new chapter discusses anal-rectal cytology and includes illustrations and specific adequacy criteria.

We, on behalf of the American Society of Cytopathology, are pleased to be a part of this ongoing process, and hope that this updated and expanded atlas will prove useful in your cytology practice. Additional information can also be found on the Bethesda educational Web site www.cytopathology.org/NIH.

Diane D. Davey, MD
Lexington, Kentucky

David C. Wilbur, MD
Boston, Massachusetts
August 2003

References

1. Solomon D, Davey DD, Kurman R, et al. The 2001 Bethesda system: terminology for reporting results of cervical cytology. *JAMA* 2002;287:2114–2119.
2. Kurman RJ, Solomon D. *The Bethesda System for Reporting Cervical/Vaginal Cytologic Diagnoses*. New York: Springer-Verlag, 1994.
3. Stoler MH, Schiffman M. Interobserver variability of cervical cytologic and histologic interpretations: realistic estimates from the ASCUS-LSIL triage study. *JAMA* 2001;285: 1500–1505.

The 2001 BETHESDA SYSTEM

SPECIMEN TYPE:

Indicate conventional smear (Pap smear) vs. liquid-based preparation vs. other

SPECIMEN ADEQUACY

❑ Satisfactory for evaluation (*describe presence or absence of endocervical/transformation zone component and any other quality indicators, e.g., partially obscuring blood, inflammation, etc.*)
❑ Unsatisfactory for evaluation . . . (*specify reason*)
    ❑ Specimen rejected/not processed (*specify reason*)
    ❑ Specimen processed and examined, but unsatisfactory for evaluation of epithelial abnormality because of (*specify reason*)

GENERAL CATEGORIZATION (*optional*)

❑ Negative for Intraepithelial Lesion or Malignancy
❑ Other: See Interpretation/Result (*e.g., endometrial cells in a woman ≥40 years of age*)
❑ Epithelial Cell Abnormality: See Interpretation/Result (*specify* 'squamous' *or* 'glandular' *as appropriate*)

INTERPRETATION/RESULT

NEGATIVE FOR INTRAEPITHELIAL LESION OR MALIGNANCY
(*when there is no cellular evidence of neoplasia, state this in the General Categorization above and/or in the Interpretation/Result section of the report--whether or not there are organisms or other non-neoplastic findings*)

ORGANISMS:

➢ *Trichomonas vaginalis*

➢ Fungal organisms morphologically consistent with *Candida* spp.

➢ Shift in flora suggestive of bacterial vaginosis

➢ Bacteria morphologically consistent with *Actinomyces* spp.

➢ Cellular changes consistent with herpes simplex virus

OTHER NON-NEOPLASTIC FINDINGS (*Optional to report; list not inclusive*):

➢ Reactive cellular changes associated with
  • inflammation (includes typical repair)
  • radiation
  • intrauterine contraceptive device (IUD)

➢ Glandular cells status posthysterectomy

➢ Atrophy

OTHER

➢ Endometrial cells (*in a woman ≥40 years of age*)

(*Specify if "negative for squamous intraepithelial lesion"*)

EPITHELIAL CELL ABNORMALITIES

SQUAMOUS CELL

➢ Atypical squamous cells
  • of undetermined significance (ASC-US)
  • cannot exclude HSIL (ASC-H)

➢ Low-grade squamous intraepithelial lesion (LSIL)
  (*encompassing: HPV/mild dysplasia/CIN 1*)

➢ High-grade squamous intraepithelial lesion (HSIL)
  (*encompassing: moderate and severe dysplasia, CIS; CIN 2 and CIN 3*)
  • with features suspicious for invasion (*if invasion is suspected*)

➢ Squamous cell carcinoma

GLANDULAR CELL

➢ Atypical
  • endocervical cells (NOS *or specify in comments*)
  • endometrial cells (NOS *or specify in comments*)
  • glandular cells (NOS *or specify in comments*)

➢ Atypical
  • endocervical cells, favor neoplastic
  • glandular cells, favor neoplastic

➢ Endocervical adenocarcinoma in situ

➢ Adenocarcinoma
  • endocervical
  • endometrial
  • extrauterine
  • not otherwise specified (NOS)

OTHER MALIGNANT NEOPLASMS: *(specify)*

ANCILLARY TESTING

Provide a brief description of the test method(s) and report the result so that it is easily understood by the clinician.

AUTOMATED REVIEW

If case examined by automated device, specify device and result.

EDUCATIONAL NOTES AND SUGGESTIONS (optional)

Suggestions should be concise and consistent with clinical follow-up guidelines published by professional organizations (references to relevant publications may be included).

Chapter 1

Specimen Adequacy

George G. Birdsong, Diane D. Davey, Teresa M. Darragh,
Paul A. Elgert, and Michael Henry

Background

Evaluation of specimen adequacy is considered by many to be the single most important quality assurance component of the Bethesda System. Earlier versions of Bethesda included three categories of adequacy: Satisfactory, Unsatisfactory, and a "borderline" category initially termed "Less than optimal" and then renamed "Satisfactory but limited by" in 1991. The 2001 Bethesda System eliminates the borderline category, in part, because of confusion among clinicians as to the appropriate follow-up for such findings and also due to the variability in reporting "Satisfactory but limited by" among laboratories.[1] To provide a clearer indication of adequacy, specimens are now designated as "Satisfactory" or "Unsatisfactory."

Previous criteria for determining adequacy were based on expert opinion and the few available studies in the literature. Laboratory implementation of some of these criteria was shown to be poorly reproducible.[2–4] In addition, the increasing use of liquid-based cytology necessitated developing criteria applicable to these preparations. The 2001 Bethesda adequacy criteria are based on published data to the extent possible and are tailored to conventional smears and liquid-based specimens.

Adequacy Categories

Satisfactory

Satisfactory for evaluation *(describe presence or absence of endocervical/ transformation zone component and any other quality indicators, e.g., partially obscuring blood, inflammation, etc.)*

Unsatisfactory

For unsatisfactory specimens, indicate whether or not the laboratory has processed/evaluated the slide. Suggested wording:

A. Rejected Specimen:
   Specimen rejected (not processed) because _____ (specimen not labeled, slide broken, etc.)
B. Fully evaluated, unsatisfactory specimen:
   Specimen processed and examined, but unsatisfactory for evaluation of epithelial abnormality because of _____ (obscuring blood, etc.)

Additional comments/recommendations, as appropriate

Explanatory Notes

For "Satisfactory" specimens, information on transformation zone sampling and other adequacy qualifiers is also included. Providing clinicians/specimen takers with regular feedback on specimen quality promotes heightened attention to specimen collection and consideration of improved sampling devices and technologies.

Any specimen with abnormal cells [atypical squamous cells of undetermined significance (ASC-US), atypical glandular cells (AGC), or worse] is by definition satisfactory for evaluation. If there is concern that the specimen is compromised, a note may be appended indicating that a more severe abnormality cannot be excluded.

Unsatisfactory specimens that are processed and evaluated require considerable time and effort. Although such specimens cannot exclude an epithelial lesion, information such as the presence of organisms, or endometrial cells in women 40 years of age or older, etc. (see Chapter 3, Endometrial Cells), may help direct further patient management.[5] Note that the presence of benign endometrial cells does not make an otherwise unsatisfactory specimen satisfactory.

A longitudinal study[6] found that unsatisfactory specimens that were processed and evaluated were more often from high-risk patients, and a significant number of these were followed by a squamous intraepithelial lesion (SIL)/cancer when compared to a cohort of satisfactory index specimens.

Minimum Squamous Cellularity Criteria

Conventional Smears (Figs. 1.1–1.5)

An adequate conventional specimen has an estimated minimum of approximately 8,000 to 12,000 well-preserved and well-visualized squamous epithelial cells. *Note: This minimum cell range should be estimated, and laboratories should not count individual cells in conventional smears.* This range applies only to squamous cells; endocervical cells and completely obscured cells should be excluded from the estimate as much as feasible. However, squamous metaplastic cells can be counted as squamous cells during cellularity assessment. This cellularity range should not be considered a rigid threshold (see comments below). "Reference images" of known cellularity are illustrated in Figures 1.1 to 1.5. These reference images have been computer edited to simulate the appearance of 4× fields on conventional smears. Cytologists should compare these images to specimens in question to determine if there are a sufficient number of fields with approximately equal or greater cellularity than the reference image. For instance, if an image corresponding to a 4× field with 1000 cells was used as the reference, a specimen would need to have at least eight such 4× fields to be deemed to have adequate cellularity.

Liquid-Based Preparations (Figs. 1.6–1.10)

An adequate liquid-based preparation (LBP) should have an estimated minimum of at least 5000 well-visualized/well-preserved squamous cells. Some have advocated that LBPs with 5,000 to 20,000 cells are of borderline or low squamous cellularity. In specimens with an apparent borderline or low squamous cellularity, an estimation of total cellularity can be obtained by performing representative field cell counts. A minimum of 10 microscopic fields, usually at 40×, should be assessed along a diameter that includes the center of the preparation and an average number of cells per field estimated. When there are holes or empty areas on the preparation, the percentage of the hypocellular areas should be estimated, and the fields counted should reflect this proportion. SurePath (TriPath Imaging, Inc., Burlington, NC) slides require higher cell density because of the smaller preparation diameter.

Table 1.1 provides the average number of cells per field required to achieve a minimum of 5000 cells on an LBP given a certain preparation diameter and eyepiece (ocular). For individuals using oculars and preparations not shown, the formula is: number of cells required per field = 5000/(area of preparation/area of field). As of this writing, the diameters of SurePath and ThinPrep (Cytyc Corporation, Boxborough, MA) preparations are 13 and 20 millimeters (mm) respectively. The diameter of a microscopic field in millimeters is the field number of the eyepiece divided by the magnification of the objective. The area of the field is then determined by the formula used to calculate the area of a circle [pi × radius squared, πr^2]. The magnification power of the ocular does not affect this calculation. See http://www.olympusmicro.com/primer/anatomy/oculars.html for additional explanation of the pertinent optical principles.

Figures 1.6 to 1.10 show cell coverage or density in satisfactory, borderline satisfactory, and unsatisfactory LBP. These are not reference images, as they do not represent an entire microscopic field; thus, the cell density shown in the images cannot be compared directly to Table 1.1 for estimation of squamous cellularity.

In some instances the cellularity on the prepared slide may not be representative of the collected sample. Slides with fewer than 5000 cells should be examined to determine if the reason for the scant cellularity is a technical problem in preparation of the slide such as excessive blood in the specimen. When a technical problem is identified and corrected, a repeat preparation may yield adequate cellularity. However, the adequacy of each slide should be determined separately and not cumulatively. Attempts to determine cellularity cumulatively by summing the cellularity of multiple inadequate slides may be confounded by uncertainty regarding the true cellularity of the specimen (not slide), which might be substantially less than in a specimen of normal cellularity. This matter is in need of more research, and this guideline may change in the future. However, given the relatively low minimum criterion for adequate cellularity, caution is warranted in borderline cases. The report should clarify whether blood, mucus, or inflammation contributed to an unsatisfactory sample, or whether the problem was simply low squamous cellularity.

Explanatory Notes

It is recognized that strict objective criteria may not be applicable to every case. Some slides with cell clustering, atrophy, or cytolysis are technically difficult to count, and there may be clinical circumstances in which a lower cell number may be considered adequate. Laboratories should apply professional judgment and employ hierarchical review when evaluating these rare borderline adequacy slides. It should also be kept in mind that the minimum cellularity criteria described were developed for use with cervical cytology specimens. In vaginal specimens (post total hysterectomy), laboratories should exercise judgment in reporting cellularity based on the

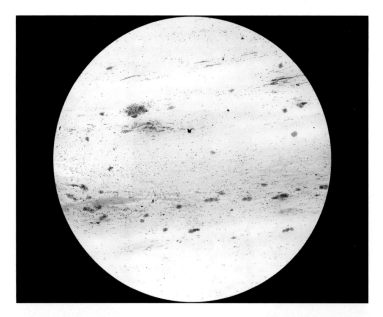

FIGURE 1.1. Squamous cellularity: This image depicts the appearance of a 4× field of a conventional Pap smear with approximately 75 cells. The specimen is unsatisfactory if all fields have this level, or less, of cellularity. It is to be used as a guide in assessing the squamous cellularity of a conventional smear. (Used with permission, © George Birdsong, 2003.)

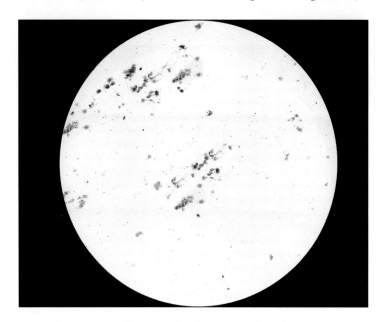

FIGURE 1.2. Squamous cellularity: This image depicts the appearance of a 4× field of a conventional Pap smear with approximately 150 cells. If all fields have this level of cellularity, the specimen will meet the minimum cellularity criterion, but by only a small margin. (Used with permission, © George Birdsong, 2003.)

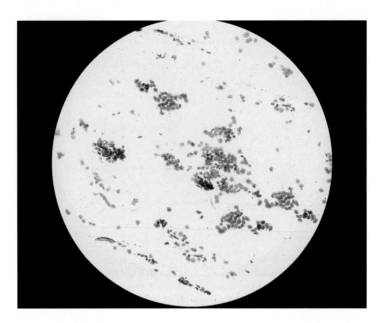

Figure 1.3. Squamous cellularity: This image depicts the appearance of a 4× field of a conventional Pap smear with approximately 500 cells. A minimum of 16 fields with similar (or greater) cellularity are needed to call the specimen adequate. (Used with permission, © George Birdsong, 2003.)

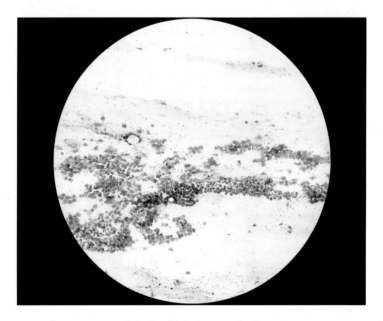

Figure 1.4. Squamous cellularity: This image depicts the appearance of a 4× field of a conventional Pap smear with approximately 1000 cells. A minimum of 8 fields with similar (or greater) cellularity are needed to call the specimen adequate. (Used with permission, © George Birdsong, 2003.)

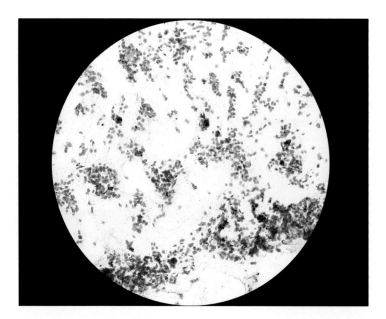

FIGURE 1.5. Squamous cellularity: This image depicts the appearance of a 4× field of a conventional Pap smear with approximately 1400 cells. A minimum of 6 fields with similar (or greater) cellularity are needed to call the specimen adequate. (Used with permission, © George Birdsong, 2003.)

clinical and screening history. A lower cellularity may be acceptable under these circumstances. Laboratories have flexibility in determining which method for cellularity estimation is best suited for their practice setting.

The recommendation for a minimum cellularity of 5000 cells for an LBP is based on preliminary scientific evidence.[7,8] This threshold is lower than the 8,000 to 12,000 minimum cellularity for conventional smears. LBPs, by virtue of the preparation methodology, present a more random (and presumably more representative) sampling of the collected cervical material as compared to conventional smears. Although there are significant differences among various LBP procedures, there are not sufficient data to justify different minimum cellularities for the LBPs currently on the market.

One preliminary study reported a higher detection rate of high-grade lesions when cellularity on LBPs exceeded 20,000.[9] However, this study did not directly investigate the possible relationship between specimen cellularity and false-negative rates, which is not necessarily the same as the relationship between specimen cellularity and the detection of high-grade lesions. Laboratories may choose to append a quality indicator com-

TABLE 1.1. GUIDELINES FOR ESTIMATING CELLULARITY OF LIQUID-BASED PREPARATIONS

Prep. diameter (mm)	Area (mm²)	FN20 eyepiece/ 10× objective		FN20 eyepiece/ 40× objective		FN22 eyepiece/ 10× objective		FN22 eyepiece/ 40× objective	
		Number of fields at FN20, 10×	Number of cells/field for 5K total	Number of fields at FN20, 40×	Number of cells/field for 5K total	Number of fields at FN22, 10×	Number of cells/field for 5K total	Number of fields at FN22, 40×	Number of cells/field for 5K total
13	132.7	42.3	118.3	676	7.4	34.9	143.2	559	9.0
20	314.2	100	50.0	1600	3.1	82.6	60.5	1322	3.8

FN, field number.

FIGURE 1.6. Unsatisfactory due to scant squamous cellularity. Endocervical cells are seen in a honeycomb arrangement [liquid-based preparation (LBP), ThinPrep, 10×].

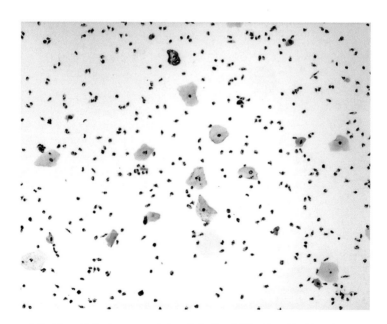

FIGURE 1.7. Unsatisfactory—scant cellularity (LBP, SurePath). Although this image cannot be directly compared to a microscopic field, this SurePath slide had fewer than 8 cells per 40× field. A SurePath specimen with this level of cellularity throughout the preparation would have fewer than 5000 cells.

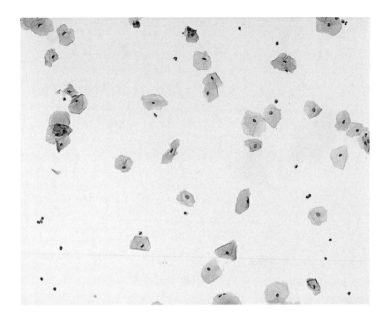

Figure 1.8. Satisfactory, but borderline squamous cellularity (LBP, SurePath). (At 40×, there were approximately 11 cells per field when 10 microscopic fields along a diameter were evaluated for squamous cellularity; this would give an estimated total cell count between 5,000 and 10,000.)

Figure 1.9. Satisfactory, but borderline squamous cellularity (LBP, Thin-Prep): 10× fields of a ThinPrep specimen should have at least this level of cellularity to be considered satisfactory. (At 40×, in this ThinPrep specimen, there were approximately 4 cells per field, which would correspond to slightly over 5000 cells. Note that this level of cell density would be unsatisfactory in a SurePath LBP [See Fig. 1.7], corresponding to less than 5,000 cells because of the smaller preparation diameter.)

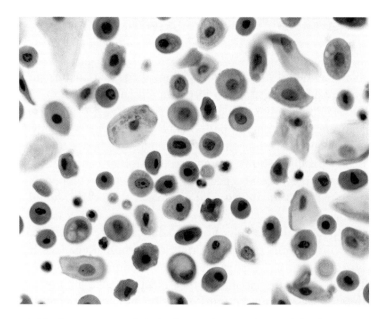

FIGURE 1.10. Squamous cellularity is satisfactory in this LBP from a 70-year-old woman with an atrophic cell pattern (LBP, SurePath). LBPs may show less nuclear enlargement than CPs (conventional preparations) due to fixation in the suspended state. The transformation zone component(s) may be difficult to assess in atrophy.

ment such as "borderline or low squamous cellularity" on such specimens that meet minimal criteria for satisfactory cellularity but have only 5,000 to 20,000 cells. Patients should be managed similarly to other patients with quality indicator statements.[5]

Cellularity can be quickly and reproducibly estimated in LBPs.[7,10] Some manufacturers include estimation of LBP cellularity during training. Preliminary studies show that reference images methodology for smears is quickly learned and has better interobserver reproducibility than the previous Bethesda 10% slide coverage criterion.[11]

Additional studies relating sensitivity to cell number would be useful for all preparation types. Guidelines may be revised in the future if studies demonstrate that squamous cellularity criteria different from those outlined are more appropriate.

Endocervical/Transformation Zone Component (Figs. 1.6, 1.11–1.16)

For both conventional smears and LBPs, an adequate transformation zone component requires at least 10 well-preserved endocervical or squamous metaplastic cells, singly or in clusters (Figs. 1.6, 1.11–1.15). The presence or absence of a transformation zone component is reported in the specimen adequacy section unless the woman has had a total hysterectomy. If the specimen shows a high-grade lesion or cancer, it is not necessary to report presence/absence of a transformation component.

Degenerated cells in mucus and parabasal-type cells should not be counted in assessing transformation zone sampling. It may be difficult to distinguish parabasal-type cells from squamous metaplastic cells in specimens showing atrophy due to a variety of hormonal changes including menopause, postpartum changes, and progestational agents (Figs. 1.10, 1.16) In such cases, the laboratory may elect to make a comment about the difficulty of assessing the transformation zone component.

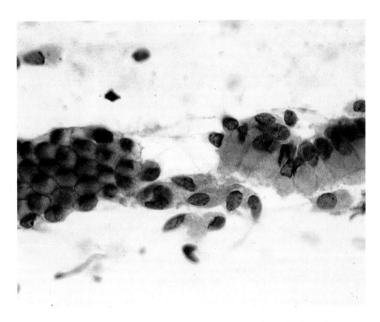

FIGURE 1.11. Endocervical cells (CP). Distinct cytoplasmic borders are seen in the cluster of cells on the left, giving a "honeycomb" appearance. The cell cluster on the right is seen from a side view, giving the "picket fence" appearance.

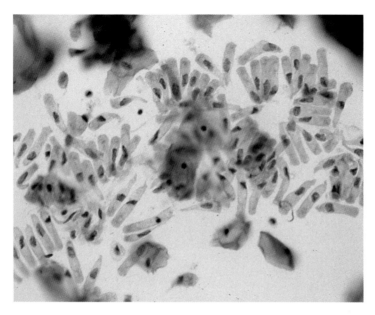

FIGURE 1.12. Endocervical cells (LBP). Dissociation is more frequent in liquid-based preparations than in conventional smear preparations.

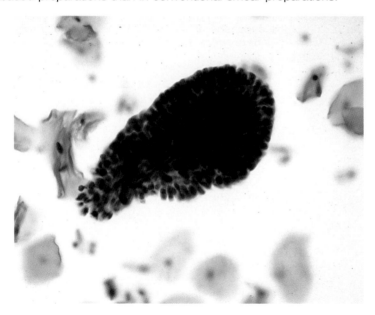

FIGURE 1.13. Endocervical cells (LBP). Routine screening, 27-year-old woman, NILM on followup. Normal endocervical cells may appear in large hyperchromatic fragments, often in the center of some LBPs. The thickness of the fragment may give the appearance of architectural disarray; however, note normal appearing cells at the periphery of the fragment. Additionally, focusing up and down through the fragment reveals normal spacing of cells, distinct cytoplasmic borders, and bland nuclear chromatin. Normal endocervical cell groups with this appearance should not be confused with dysplastic/neoplastic clusters that show more crowding (even within a single layer of cells), nuclear enlargement, nuclear membrane irregularity, and abnormal chromatin pattern.

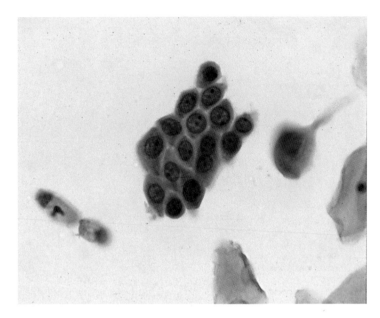

[*] FIGURE 1.14. Endocervical cells (LBP). Normal endocervical cells from the upper region of the endocervical canal can mimic squamous metaplastic cells. [[*] Bethesda Interobserver Reproducibility Project (BIRP) image (see xviii Introduction).]

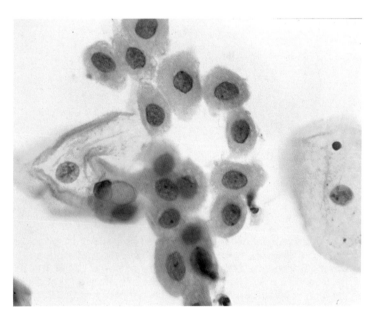

[*] FIGURE 1.15. Normal squamous metaplastic cells (LBP). Routine screening, 28-year-old woman.

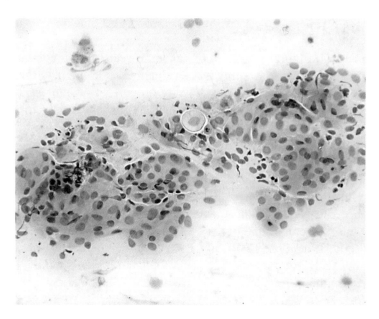

FIGURE 1.16. Atrophy (CP).

Explanatory Notes

Data on the importance of the endocervical/transformation zone (EC/TZ) component are conflicting. Cross-sectional studies show that SIL cells are more likely to be present on specimens in which EC/TZ cells are present.[12–14] However, retrospective cohort studies have shown that women with specimens that lack EC/TZ elements are not more likely to have squamous lesions on follow-up than are women with EC/TZ.[15–18] Birdsong recently reviewed this subject.[19] A recent study that included colposcopic evaluation of all women with abnormal liquid-based cytology or human papillomavirus (HPV) results plus a random sample of those with negative test results failed to show an association between absent EC/TZ component and missed high-grade lesions.[20] Finally, retrospective case-control studies have failed to show an association between false-negative interpretations of specimens and lack of EC.[21,22]

The implications of the EC/TZ component could change in the future as the incidence of endocervical adenocarcinoma is increasing.[23–25] The relationship between the detection of adenocarcinoma and the presence of endocervical cells on cervical cytology specimens is unexplored as of this writing.

Obscuring Factors (Figs. 1.17, 1.18)

Specimens with more than 75% of squamous cells obscured should be termed unsatisfactory, assuming that no abnormal cells are identified (Fig. 1.17). When 50% to 75% of the cells are obscured, a statement describing the specimen as partially obscured should follow the satisfactory term. The percentage of cells obscured, not the slide area obscured, should be evaluated, although minimal cellularity criteria should also be applied. Nuclear preservation and visualization are of key importance, and changes such as cytolysis and partial obscuring of cytoplasmic detail may not necessarily interfere with specimen evaluation. Abundent cytolysis may be mentioned as a quality indicator, but most such specimens do not qualify as "unsatisfactory" unless nearly all nuclei are devoid of cytoplasm. Similar criteria apply to LBPs. In LBPs with some obscuring factors and borderline cellularity (see Figs. 1.8, 1.9), laboratories should estimate whether minimum numbers of well-visualized squamous cells are present as described above. When particular cells or areas of diagnostic in-

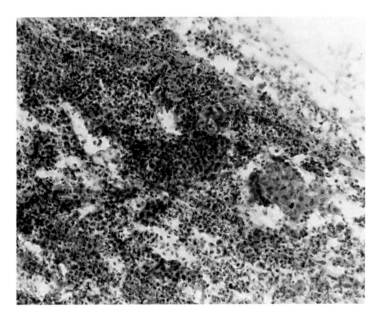

FIGURE 1.17. Unsatisfactory due to obscuring white blood cells (CP). If 50%–75% of the epithelial cells are covered, obscuring inflammation should be mentioned in the quality indicators section of the report (>75% obscuring is considered unsatisfactory if no abnormal cells are identified). In assessing the adequacy of a slide with respect to obscuring factors and cellularity, one should keep in mind that the minimum cellularity criteria refer to *well-visualized* cells.

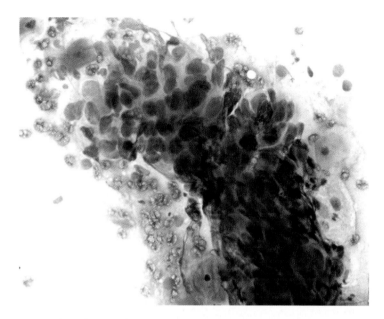

Figure 1.18. Satisfactory for evaluation; extensive air drying artifact present. Atypical squamous cells cannot exclude high grade squamous intraepithelial lesion (ASC-H) (CP). Enlarged, pale nuclei with indistinct chromatin. The nuclei are crowded and lack an orderly architectural arrangement. Note that if the interpretation is atypical cells or worse, then the specimen cannot be considered "unsatisfactory" regardless of specimen quality. Follow-up in this case was CIN 2.

terest are obscured, a report comment can be added: e.g., "air-drying of possible atypical cells" (Fig. 1.18).

Explanatory Notes

Specimens with partial obscuring factors have been shown to have fair interobserver reproducibility of adequacy assessment.[26] Although retrospective case-control studies fail to show that partial obscuring factors indicate risk for a false-negative report,[21,22] prospective studies have not been done. Reporting obscuring factors may be indicated because of patient care or quality concerns.

Management

Information on adequacy and any implications for patient follow-up may be provided optionally in an educational note. The American Society for Colposcopy and Cervical Pathology has published management guidelines for specimen adequacy and quality indicators based on the 2001 Bethesda terminology.[5]

Sample Reports

Example 1:

Satisfactory for evaluation; endocervical/transformation zone component present.

Interpretation:

Negative for intraepithelial lesion or malignancy.

Example 2:

Satisfactory for evaluation; endocervical/transformation zone component absent/insufficient.

Interpretation:

Negative for intraepithelial lesion or malignancy.

Optional Note:

Data are conflicting regarding the significance of endocervical/transformation zone elements. A repeat cervical cytology in 12 months is generally suggested. (ASCCP Patient Management Guidelines: Am J Clin Pathol 2002;118:714–718)

Example 3:

Unsatisfactory for evaluation; Specimen processed and examined, but unsatisfactory for evaluation of epithelial abnormality because of obscuring inflammation.

Comment:

Trichomonas vaginalis identified.

Consider repeat cervical cytology/Pap test after treatment of *Trichomonas.*

Example 4:

Interpretation:

Specimen processed and examined, but unsatisfactory for evaluation of epithelial abnormality because of insufficient squamous cellularity. Partially obscuring blood identified.

Optional:

Unsatisfactory for evaluation.

Comment:

Endometrial cells present consistent with day 5 of LMP (last menstrual period) as provided.

Example 5:

Unsatisfactory for evaluation; Specimen rejected because slide was received unlabeled.

Bethesda System 2001 Workshop Forum Group Moderators:

Diane D. Davey, M.D., George Birdsong, M.D., Henry W. Buck, M.D., Teresa Darragh, M.D., Paul Elgert, C.T. (ASCP), Michael Henry, M.D., Heather Mitchell, M.D., Suzanne Selvaggi, M.D.

References

1. Davey DD, Woodhouse S, Styer PE, et al. Atypical epithelial cells and specimen adequacy: current laboratory practices of participants in the College of American Pathologists Interlaboratory Comparison Program in Cervicovaginal Cytology. *Arch Pathol Lab Med* 2000;124:203–211.
2. Gill GW. Pap smear cellular adequacy: what does 10% coverage look like? What does it mean? *Acta Cytol* 2000;44:873 (abstract).
3. Renshaw AA, Friedman MM, Rahemtulla A, et al. Accuracy and reproducibility of estimating the adequacy of the squamous component of cervicovaginal smears. *Am J Clin Pathol* 1999;111:38–42.
4. Valente PT, Schantz HD, Trabal JF. The determination of Papanicolaou smear adequacy using a semiquantitative method to evaluate cellularity. *Diagn Cytopathol* 1991;7:576–580.
5. Davey DD, Austin RM, Birdsong G, et al. ASCCP Patient Management Guidelines: Pap test specimen adequacy and quality indicators. *J Lower Genital Tract Dis* 2002;6:195–199. (Also published in *Am J Clin Pathol* 2002;118:714–718.)
6. Ransdell JS, Davey DD, Zaleski S. Clinicopathologic correlation of the unsatisfactory Papanicolaou smear. *Cancer (Cancer Cytopathol)* 1997;81:139–143.
7. Geyer JW, Carrico C, Bishop JW. Cellular constitution of Autocyte PREP cervicovaginal samples with biopsy-confirmed HSIL. *Acta Cytol* 2000;44:505 (abstract).
8. Studeman KD, Ioffe OB, Puszkiewicz J, et al. The effect of cellularity on the sensitivity of squamous lesions in liquid-based cervical cytology. *Acta Cytol* 2003;47:605–610.
9. Bolick DR, Kerr J, Staley BE, et al. Effect of cellularity in the detection rates of high grade and low grade squamous intraepithelial lesions. *Acta Cytol* 2002;46:922–923 (abstract).
10. Haroon S, Samayoa L, Witzke D, Davey D. Reproducibility of cervicovaginal Thin-Prep cellularity assessment. *Diagn Cytopathol* 2002;26:19–21.
11. Sheffield MV, Simsir A, Talley L, et al. Interobserver variability in assessing adequacy of the squamous component in conventional cervicovaginal smears. *Am J Clin Pathol* 2003;119:367–373.
12. Martin-Hirsch P, Lilford R, Jarvis G, et al. Efficacy of cervical-smear collection devices: a systematic review and meta-analysis. *Lancet* 1999;354:1763–1770.
13. Vooijs PG, Elias A, van der Graaf Y, et al. Relationship between the diagnosis of epithelial abnormalities and the composition of cervical smears. *Acta Cytol* 1985;29:323–328.
14. Mintzer MP, Curtis P, Resnick JC, et al. The effect of the quality of Papanicolaou smears on the detection of cytologic abnormalities. *Cancer (Cancer Cytopathol)* 1999;87:113–117.

15. Bos AB, van Ballegooijen M, van den Akker-van Marle E, et al. Endocervical status is not predictive of the incidence of cervical cancer in the years after negative smears. *Am J Clin Pathol* 2001;115:851–855.
16. Kivlahan C, Ingram E. Papanicolaou smears without endocervical cells. Are they inadequate? *Acta Cytol* 1986;30:258–260.
17. Mitchell HS. Longitudinal analysis of histologic high-grade disease after negative cervical cytology according to endocervical status. *Cancer (Cancer Cytopathol)* 2001; 93:237–240.
18. Mitchell H, Medley G. Longitudinal study of women with negative cervical smears according to endocervical status. *Lancet* 1991;337:265–267.
19. Birdsong GG. Pap smear adequacy: is our understanding satisfactory . . . or limited? *Diagn Cytopathol* 2001;24:79–81.
20. Baer A, Kiviat NB, Kulasingam S, et al. Liquid-based Papanicolaou smears without a transformation zone component: should clinicians worry? *Obstet Gynecol* 2002;99: 1053–1059.
21. Mitchell H, Medley G. Differences between Papanicolaou smears with correct and incorrect diagnoses. *Cytopathology* 1995;6:368–375.
22. O'Sullivan JP, A'Hern RP, Chapman PA, et al. A case-control study of true-positive versus false-negative cervical smears in women with cervical intraepithelial neoplasia (CIN) III. *Cytopathology* 1998;9:155–161.
23. Alfsen GC, Thoresen SO, Kristensen GB, et al. Histopathologic subtyping of cervical adenocarcinoma reveals increasing incidence rates of endometrioid tumors in all age groups: a population based study with review of all nonsquamous cervical carcinomas in Norway from 1966 to 1970, 1976 to 1980, and 1986 to 1990. *Cancer* 2000; 89:1291–1299.
24. Stockton D, Cooper P, Lonsdale RN. Changing incidence of invasive adenocarcinoma of the uterine cervix in East Anglia. *J Med Screen* 1997;4:40–43.
25. Zheng T, Holford TR, Ma Z, et al. The continuing increase in adenocarcinoma of the uterine cervix: a birth cohort phenomenon. *Int J Epidemiol* 1996;25:252–258.
26. Spires SE, Banks ER, Weeks JA, et al. Assessment of cervicovaginal smear adequacy. The Bethesda system guidelines and reproducibility. *Am J Clin Pathol* 1994;102: 354–359.

Chapter 2

Non-Neoplastic Findings

Nancy A. Young, Marluce Bibbo, Sally-Beth Buckner,
Terence J. Colgan, and Marianne U. Prey

Negative for Intraepithelial Lesion or Malignancy

(when there is no cellular evidence of neoplasia, state this in the General Categorization and/or in the Interpretation/Result section of the report-- whether or not there are organisms or other non-neoplastic findings)

Organisms:

- ❏ *Trichomonas vaginalis*
- ❏ Fungal organisms morphologically consistent with *Candida* spp.
- ❏ Shift in flora suggestive of bacterial vaginosis
- ❏ Bacteria morphologically consistent with *Actinomyces* spp.
- ❏ Cellular changes consistent with herpes simplex virus

Other Non-Neoplastic Findings

(optional to report; list not inclusive):

- ❏ Reactive cellular changes associated with:
  - ➢ inflammation (includes typical repair)
  - ➢ radiation
  - ➢ intrauterine contraceptive device (IUD)
- ❏ Glandular cells status post hysterectomy
- ❏ Atrophy

Background

In previous versions of The Bethesda System (TBS), infections and reactive cellular changes were reported under the categorical heading of "Benign Cellular Changes" (BCC); but under the General Categorization section, BCC was separate from "Within normal limits" (WNL). Therefore, some clinicians viewed BCC as something other than WNL and felt obligated to do more than routine screening for women with a BCC designation. To emphasize the 'negative' nature of an interpretation of reactive changes, the 2001 Bethesda System collapses the BCC and WNL categories into a single category: "Negative for Intraepithelial Lesion or Malignancy" (NILM). This term is used both as the general categorization and/or as the interpretation in the report.

The category of "Infections" has been changed to "Organisms" because the presence of some organisms reflects colonization rather than clinical infection. Although not the main focus of cervical screening, providing such information may be clinically relevant in certain circumstances. Clinicians and laboratories should communicate with one another about their expectations for reporting organisms. In the absence of specific communication regarding this issue, the organisms listed should generally be reported if identified.

Cervical cytology is a screening test primarily for detection of squamous cell carcinoma of the cervix and its precursors. Criteria for reactive cellular changes are not always well defined, and consequently the interpretation may lack reproducibility.[1–6] Except for organisms discussed above, reporting non-neoplastic findings is optional and at the discretion of the laboratory. However, reasons for continuing to report non-neoplastic findings in a cervical cytology report include:

1. Utility as a triage tool and as documentation for laboratory regulations regarding referral for hierarchical review.
2. Fostering a discipline in applying cytomorphologic criteria during screening and sign-out.
3. Documentation of morphologic findings to explain differences in interpretation on review.[7]
4. Facilitation of clinical–cytologic correlation. For example, the cytologic finding of hyperkeratosis and parakeratosis may correlate with the colposcopist's assessment of the uterine cervix.
5. Documentation of reactive cellular changes in the report to spot trends in a series of cervical cytology specimens from one woman. Studies have reported a slight increase in the incidence of squamous intraepithelial lesion (SIL) in cases interpreted as reactive compared to those interpreted as within normal limits.[8,9] Reporting non-neoplastic findings may facilitate future studies to identify specific morphologic findings that better correlate with risk.

Note that the list of non-neoplastic findings in Bethesda 2001 is not comprehensive. In addition, the interpretive categories do not necessarily correspond to regulatory requirements for hierarchical supervisory review;

within the parameters of government regulation, it is up to the laboratory to specify findings that trigger such review.

Negative for Intraepithelial Lesion or Malignancy (NILM)

Specimens for which no epithelial abnormality is identified are reported as "Negative for intraepithelial lesion or malignancy" (NILM). If non-neoplastic findings are reported, NILM should still be included as an interpretation or as the General Categorization to avoid ambiguity.

Organisms

Trichomonas vaginalis (Figs. 2.1–2.3)

Criteria

Pear-shaped, oval, or round cyanophilic organism ranging in size from 15 to 30 μm (Fig. 2.1).
Nucleus is pale, vesicular, and eccentrically located.
Eosinophilic cytoplasmic granules are often evident.
Flagella are usually not seen.
Leptothrix may be seen in association with *T. vaginalis* (Fig. 2.2).

Liquid-Based Preparations

Organisms tend to be smaller due to rounding.
Occasional kite-shaped forms may be seen.
Nuclei and cytoplasmic eosinophilic granules are often better visualized.
Flagella may be preserved and identified in liquid-based preparations (LBP) (Fig. 2.3).

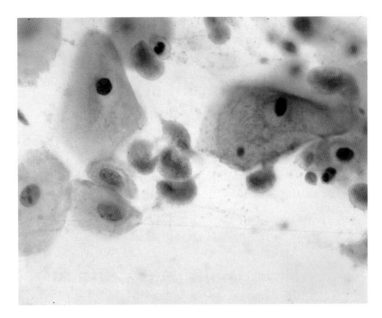

FIGURE 2.1. *Trichomonas vaginalis* (conventional preparation, CP). Pear-shaped organism with eccentrically located nucleus and eosinophilic cytoplasmic granules.

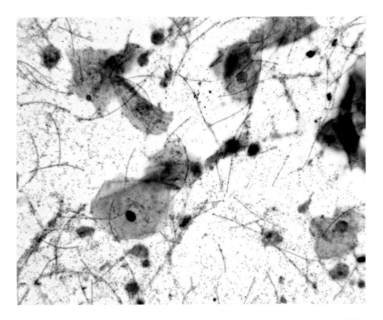

FIGURE 2.2. *Trichomonas vaginalis* in association with *Leptothrix* (CP). *Leptothrix* may be seen in association with *T. vaginalis*; this finding alone is not sufficient, but suggests the presence of trichomonads.

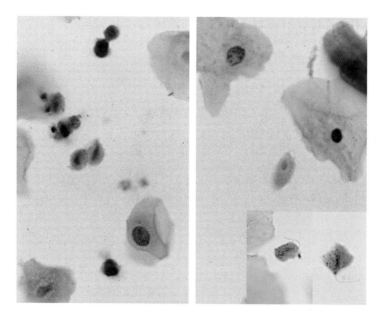

FIGURE 2.3. *Trichomonas vaginalis* (liquid-based preparation, LBP). 32-year-old woman with vaginal discharge. The organism's nucleus, cytoplasmic granules, and flagella may be better visualized on LBPs. Note the kite shape and flagella (*bottom right insert*).

Fungal Organisms Morphologically Consistent with *Candida* species (Figs. 2.4–2.6)

Criteria

Budding yeasts (3–7 μm); pseudohyphae are eosinophilic to gray-brown on the Papanicolaou stain.

Pseudohyphae, formed by elongated budding, show constrictions along their length (Fig. 2.4).

Fragmented leukocyte nuclei and rouleau formation of squamous epithelial cells "speared" by hyphae may be seen.

Liquid-Based Preparations

"Spearing" of epithelial cells is more common in LBPs and can be seen at low power even if the pseudohyphae are not prominent ("shish kebab" effect) (Fig. 2.5).

Note: Candida (Torulopsis) glabrata consists of small, uniform, round budding yeast forms surrounded by clear halos on Papanicolaou stain. Unlike other *Candida* species, it does not form pseudohyphae in vivo or in culture (Fig. 2.6).

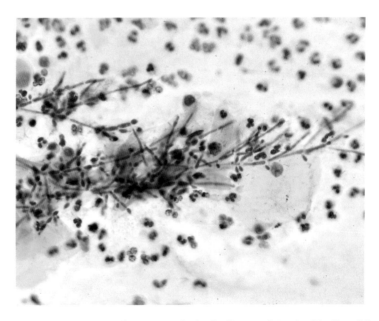

FIGURE 2.4. Fungal organisms morphologically consistent with *Candida* spp. (CP). Note pseudohyphae.

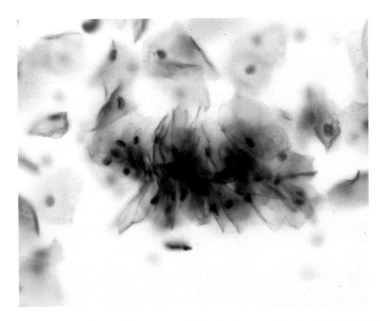

FIGURE 2.5. Fungal organisms morphologically consistent with *Candida* spp. (LBP). 45-year-old woman. Note "spearing" of squamous cells, a feature that is readily appreciated at low power, even when the pseudohyphae are not prominent. Follow up was NILM.

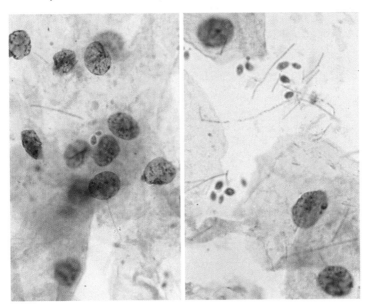

FIGURE 2.6. Fungal organisms morphologically consistent with *Candida* spp. (*Torulopsis*) *glabrata* (CP). Routine screening of 63-year-old woman. Note clear halos surrounding the yeast forms (*left*). Bacteria, not pseudohyphae, are also seen in the background.

Shift in Flora Suggestive of Bacterial Vaginosis (Figs. 2.7–2.8)

Criteria

Filmy background of small coccobacilli is evident (Fig. 2.7).
Individual squamous cells may be covered by a layer of bacteria that obscures the cell membrane, forming so-called clue cells (Fig. 2.8).
There is a conspicuous absence of lactobacilli (Fig. 2.9).

Liquid-Based Preparations

Squamous cells are covered with coccobacilli; however, the background is clean (Fig. 2.8).

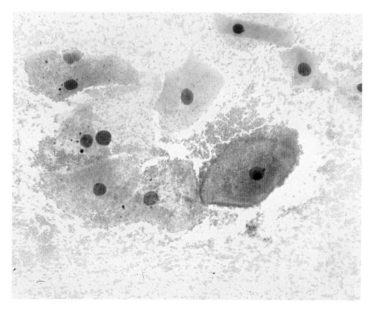

FIGURE 2.7. Shift in flora suggestive of bacterial vaginosis (CP). Note the clue cell and filmy background due to the coccobacilli.

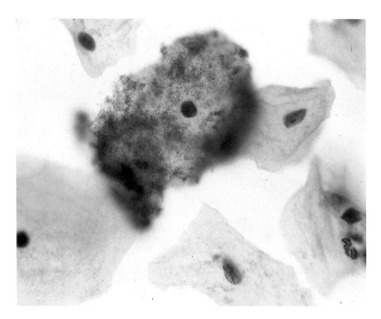

FIGURE 2.8. Shift in flora suggestive of bacterial vaginosis (LBP). 25-year-old woman. Note clue cell and the relatively clean background compared to that in CPs (Fig. 2.7).

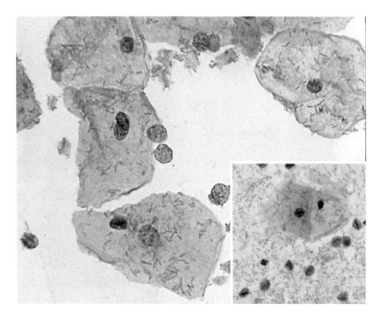

FIGURE 2.9. Lactobacilli (*left*, LBP; *bottom right insert*, CP). They are typically seen on the cell surfaces in LBP and not dispersed in the background as in CP. Compare to coccobacilli in Figs. 2.7 and 2.8.

Bacteria Morphologically Consistent with *Actinomyces* (Figs. 2.10, 2.11)

Criteria

Tangled clumps of filamentous organisms, often with acute angle branching, are recognizable as "cotton ball" clusters on low power (Fig. 2.10).

Filaments sometimes have a radial distribution or have an irregular "woolly body" appearance (Fig. 2.11).

Masses of leukocytes adherent to microcolonies of the organism, with swollen filaments or "clubs" at the periphery, may be identified.

An acute inflammatory response with polymorphonuclear leukocytes is often present.

Cellular Changes Consistent with Herpes Simplex Virus (Fig. 2.12)

Criteria

Nuclei have a "ground-glass" appearance due to intranuclear viral particles and enhancement of the nuclear envelope caused by peripheral margination of chromatin.

Dense eosinophilic intranuclear inclusions surrounded by a halo or clear zone are variably present.

Large multinucleated epithelial cells with molded nuclei are characteristic but may not always be present; mononucleate cells with the nuclear features described above may be the only finding.

Explanatory Notes

Occasionally degenerated fragments of cytoplasm or macrophages can be mistaken for trichomonads, particularly in liquid-based preparations. Therefore, at least one of the following—good nuclear detail, eosinophilic cytoplasmic granules, or flagella—should be present to make an interpretation of *Trichomonas*. When *Leptothrix* is seen, one should search for the possible presence of trichomonads.

Lactobacillus spp. constitute a major component of the normal vaginal flora (see Fig. 2.9). Predominance of coccobacilli represents a shift in vaginal flora from lactobacilli to a polymicrobial process involving several types of obligate and facultative anaerobic bacteria, including but not limited to *Gardnerella vaginalis* and *Mobiluncus* spp.[10,11] This shift in flora, with or without accompanying clue cells, is not sufficient for the

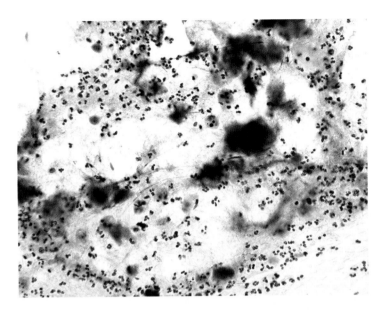

[*] FIGURE 2.10. Bacteria morphologically consistent with *Actinomyces* (CP). 41-year-old woman. Low power shows "cotton ball" appearance of tangled clumps of filamentous organisms. An acute inflammatory response is also apparent. [[*] Bethesda Interobserver Reproducibility Project (BIRP) image (see xviii Introduction).]

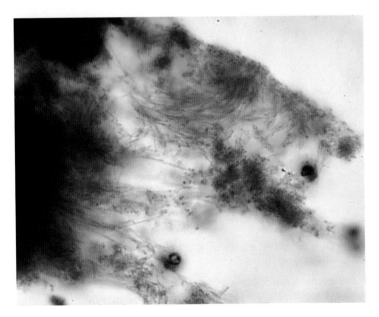

FIGURE 2.11. Bacteria morphologically consistent with *Actinomyces* (CP). Thin filamentous bacilli are seen longitudinally and on end under high magnification.

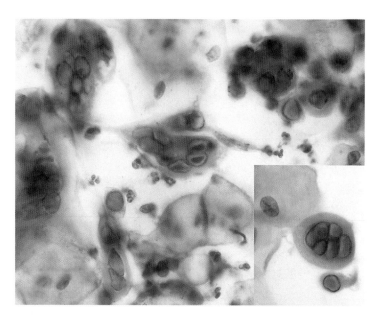

[*] FIGURE **2.12.** Cellular changes consistent with herpes simplex virus (CP). Note the eosinophilic intranuclear "Cowdry-type" inclusions. The "ground-glass" appearance of the nuclei is due to accumulation of viral particles leading to peripheral margination of chromatin (LBP, *lower right insert*).

clinical diagnosis of bacterial vaginosis because specimens obtained from any single site are not necessarily representative of the entire flora of the cervix and vagina.[12] However, the presence of coccobacilli and absence of lactobacilli do correlate with gram-stained smears of vaginal secretions and in the proper clinical context provide supportive evidence for the clinical diagnosis of bacterial vaginosis.[13] Bacterial vaginosis has been associated with pelvic inflammatory disease, preterm birth, postoperative gynecologic infections, and abnormal Pap tests.[14,15]

The presence of *Actinomyces* species in cervical cytology is associated with intrauterine contraceptive device (IUD) usage. Detection of *Actinomyces* in cervical cytology specimens along with clinical evidence of pelvic infection can help alert clinicians to the possibility of a pelvic actinomycotic abscess.[16]

Because of the controversy regarding the sensitivity and reproducibility of a cytologic finding of *Chlamydia* and the availability of more specific detection methods such as culture, enzyme-linked immunoassay, and polymerase chain reaction (PCR), the interpretation of *Chlamydia* spp. is not included in TBS.[17,18]

Other Non-Neoplastic Findings

Reactive Cellular Changes

Definition

Reactive cellular changes that are benign in nature, associated with inflammation, radiation, an IUD, or other nonspecific causes.

Reactive Cellular Changes Associated with Inflammation (Includes Typical Repair) (Figs. 2.13–2.22)

Criteria

Nuclear enlargement (one and one-half to two times the area of a normal intermediate squamous cell nucleus or more) (Figs. 2.13, 2.14, 2.21).

Endocervical cells may show greater nuclear enlargement (Figs. 2.17, 2.18).

Occasional binucleation or multinucleation may be observed.

Nuclear outlines are smooth, round, and uniform.

Nuclei may appear vesicular and hypochromatic (Figs. 2.13, 2.14).

Mild hyperchromasia may be present, but the chromatin structure and distribution remain uniformly finely granular.

Prominent single or multiple nucleoli may be present.

Cytoplasm may show polychromasia, vacuolization, or perinuclear halos but without peripheral thickening (Fig. 2.14).

Similar changes may be seen in squamous metaplastic cells (Fig. 2.15); cytoplasmic processes (spider cells) may also be seen (Fig. 2.16).

In typical repair, any of the above cellular changes may be seen; however, cells occur in flat, monolayer sheets with distinct cytoplasmic outlines (in contrast to the syncytial appearance of some high-grade lesions and cancers), streaming nuclear polarity, and typical mitotic figures. Single cells with nuclear changes are not usually seen (Figs. 2.17–2.20).

Liquid-Based Preparations (Figs. 2.21, 2.22)

Reparative groups are more rounded, with less streaming (Fig. 2.22).

Nucleoli may be more prominent.

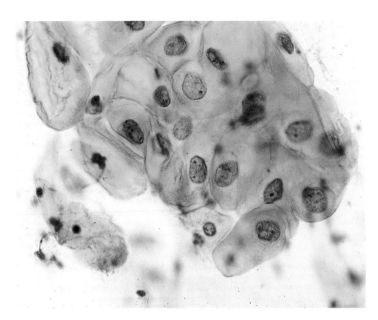

[*] **Figure 2.13.** Reactive squamous cells (CP). Mild nuclear enlargement without any significant chromatin abnormalities. (Reprinted with permission from Kurman, RJ (ed.), *Blaustein's Pathology of the Female Genital Tract*, Fourth Edition, Springer-Verlag New York, © 1994.)

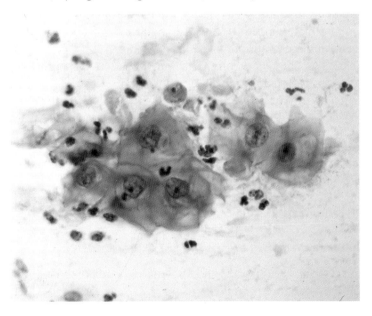

[*] **Figure 2.14.** Reactive squamous cells (CP). 26-year-old woman, day 14 of menstrual cycle with mild vaginal discharge. Squamous cells show mild nuclear enlargement, perinuclear halos, and cytoplasmic polychromasia resulting in a "moth-eaten" appearance. Trichomonads are seen in the background. Follow up was NILM.

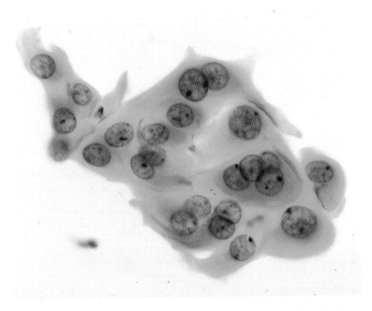

FIGURE 2.15. Squamous metaplastic cells (CP). Metaplastic cells have a higher nuclear/cytoplasmic (N:C) ratio than mature cells, but nuclear membranes are smooth and chromatin is finely granular and evenly distributed. Small round nucleoli can be seen.

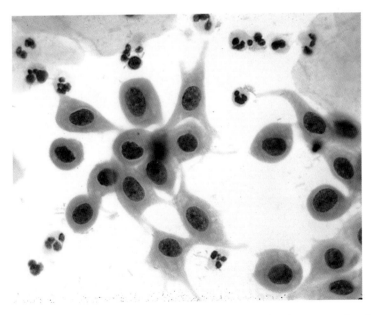

[*] **FIGURE 2.16.** Squamous metaplastic cells (CP). Routine screening from 27-year-old woman, day 8 of menstrual cycle. Note the "spidery" cytoplasmic processes, a feature that may be seen in conventional smears. Follow up was NILM.

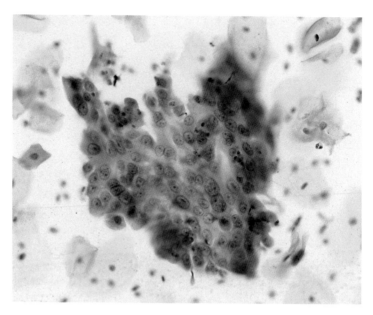

[*] **Figure 2.17.** Reactive endocervical cells (LBP). 32-year-old woman. Variation in nuclear size, prominent nucleoli, and rare intracytoplasmic polymorphonuclear leukocytes are seen; features of endocervical repair. Follow up was NILM.

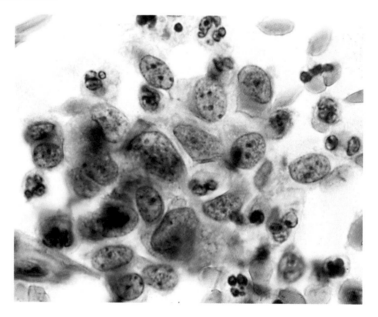

Figure 2.18. Reactive endocervical cells (CP). A 22-year-old woman status post loop electrosurgical excision procedure (LEEP) 6 months earlier for cervical intraepithelial neoplasia (CIN). Endocervical cells show variable increase in nuclear size, prominent nucleoli, and fine chromatin. Concurrent biopsy was benign.

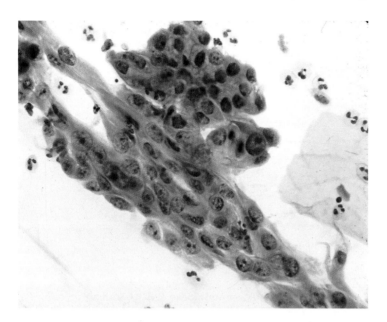

[∗] **FIGURE 2.19.** Reactive squamous cells, repair (CP). A 67-year-old woman with uterine prolapse. Flat, monolayer sheet of reparative cells with distinct cytoplasmic borders, streaming nuclear polarity, and a prominent nucleolus in almost every cell. Reactive group of endocervical cells at top center.

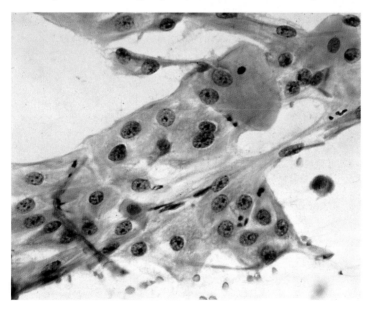

[∗] **FIGURE 2.20.** Reactive squamous cells (CP). Typical repair—squamous cells in a flat, monolayer sheet with maintenance of nuclear polarity.

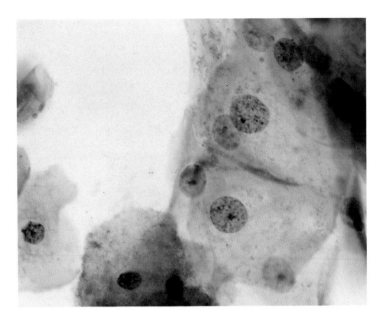

[*] FIGURE 2.21. Reactive squamous cells (LBP). Routine screen of 32-year-old woman. Although there is nuclear enlargement in the cells on the *right side*, the smooth nuclear contours and finely distributed chromatin favor reactive change over ASC-US.

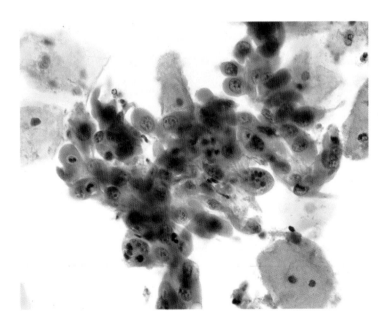

FIGURE 2.22. Repair (LBP). 32-year-old woman. Changes are similar to those seen on CPs, but cell streaming may be less apparent due to rounding of cell clusters. Note the intracytoplasmic polymorphonuclear leukocytes, another feature seen in repair. Compare to Figures 2.19 and 2.20.

Reactive Cellular Changes Associated with Radiation (Fig. 2.23)

Criteria

Cell size is markedly increased without a substantial increase in the nuclear to cytoplasmic ratio.

Bizarre cell shapes may occur.

Enlarged nuclei may show degenerative changes including nuclear pallor, wrinkling or smudging of the chromatin, and nuclear vacuolization.

Nuclei may vary in size, with some cell groups having both enlarged and normal-sized nuclei; binucleation or multinucleation is common. Mild nuclear hyperchromasia may be present.

Prominent single or multiple nucleoli may be seen if coexisting repair is present.

Cytoplasmic vacuolization and/or cytoplasmic polychromatic staining may be seen.

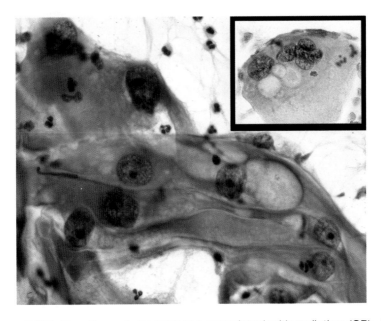

FIGURE 2.23. Reactive cellular changes associated with radiation (CP). A 40-year-old woman with history of squamous cell carcinoma of the cervix who completed radiation therapy 8 weeks earlier. Cells with enlarged nuclei, abundant vacuolated polychromatic cytoplasm, mild nuclear hyperchromasia without coarse chromatin, and prominent nucleoli. Note multinucleation (*upper right corner insert*).

Reactive Cellular Changes Associated with Intrauterine Contraceptive Device (Figs. 2.24, 2.25)

Criteria

Glandular cells may be present singly or in clusters, usually of 5 to 15 cells, amid a clean background (Fig. 2.24).

The amount of cytoplasm varies, and frequently large vacuoles may displace the nucleus, creating a signet-ring appearance (Fig. 2.24).

Occasional single epithelial cells with increased nuclear size and high nuclear/cytoplasmic ratio may be present (Fig. 2.25).

Nuclear degeneration frequently is evident.

Nucleoli may be prominent.

Calcifications resembling psammoma bodies are variably present.

Glandular Cells Status Posthysterectomy (Fig. 2.26)

Criteria

Benign-appearing endocervical-type glandular cells that cannot be differentiated from those sampled from the endocervix.

Goblet cell or mucinous metaplasia may be seen.

Round to cuboidal cells may resemble endometrial-type cells.

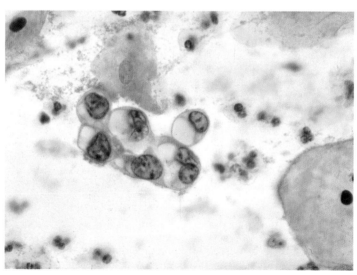

FIGURE 2.24. Reactive cellular changes associated with intrauterine contraceptive device (IUD) (CP). Note small cluster of glandular cells with cytoplasmic vacuoles displacing nuclei.

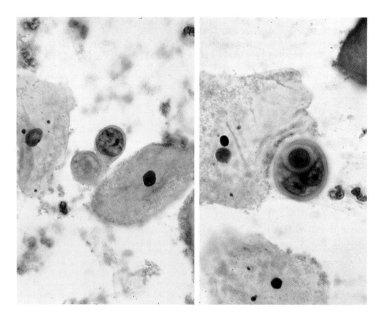

FIGURE 2.25. Reactive cellular changes associated with IUD (CP). Epithelial cells with a high nuclear/cytoplasmic ratio may mimic high-grade squamous intraepithelial lesion (HSIL) (*left*); however, the morphologic spectrum of abnormalities usually present with squamous intraepithelial lesions is absent. Presence of nucleoli in isolated cells with a high N/C ratio (*right*) is not typical of HSIL.

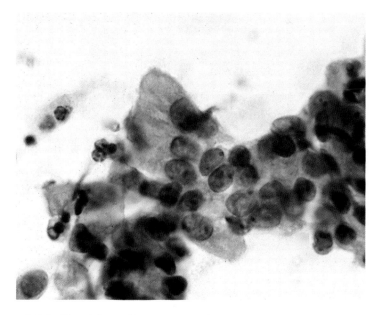

FIGURE 2.26. Glandular cells status posthysterectomy (CP). Vaginal smear from a 49-year-old woman status posthysterectomy for squamous cell cancer of the cervix, showing benign-appearing endocervical cells.

Atrophy With or Without Inflammation (Figs. 2.27–2.32)

Criteria

Flat, monolayer sheets of parabasal-like cells with preserved nuclear polarity (Fig. 2.28).

Dispersed parabasal-type cells may predominate.

Generalized nuclear enlargement, up to three to five times the area of an intermediate cell nucleus, may occur with a slight increase in nuclear/cytoplasmic ratio.

Intermediate cells tend to be normochromatic, but parabasal-type cells may have mild hyperchromasia and tend to have more elongated nuclei.

Chromatin is uniformly distributed.

Autolysis may result in naked nuclei.

An abundant inflammatory exudate and basophilic granular background that resembles tumor diathesis may be present (Fig. 2.29).

Globular collections of basophilic amorphous material (blue blobs) reflect either degenerated parabasal cells or inspissated mucus.

Degenerated orangeophilic or eosinophilic parabasal cells with nuclear pyknosis resembling "parakeratotic" cells may be present (Fig. 2.31).

Histiocytes varying in size and shape, and containing multiple, round to epithelioid nuclei and foamy or dense cytoplasm, may be seen (Fig. 2.30).

Liquid-Based Preparations

Liquid-based preparations have less nuclear enlargement than conventional smears due to immediate fixation.

Naked nuclei from autolysis may be reduced in number.

Granular background material tends to clump rather than be dispersed, yielding a "cleaner" background (Fig. 2.31); however, the clumps may "cling" to the cells and make it difficult to visualize individual cells (Fig. 2.32).

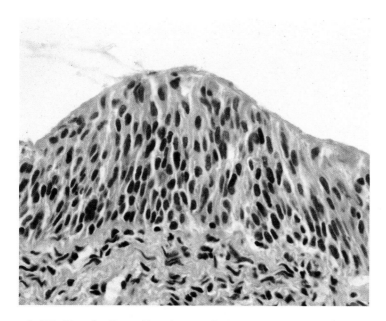

FIGURE 2.27. Atrophy "transitional metaplasia" (histology, H&E).

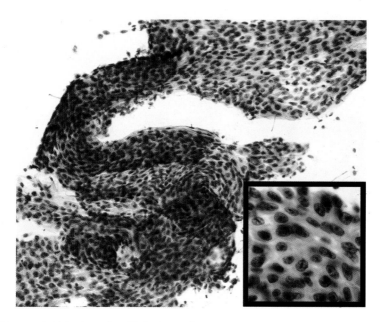

[∗] **FIGURE 2.28.** Atrophy (CP). Note flat, monolayer sheet of parabasal-type cells, with preserved nuclear polarity. Atrophic cells may have nucleoli (*lower right insert*).

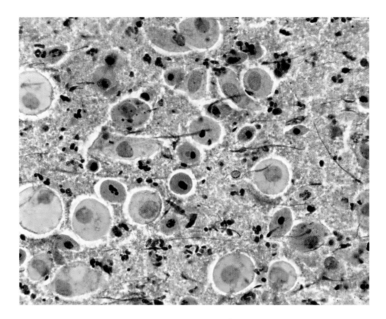

[*] FIGURE 2.29. Atrophy with inflammation ("atrophic vaginitis") (CP). Note granular debris in background, degenerating parabasal cells, and polymorphonuclear leukocytes.

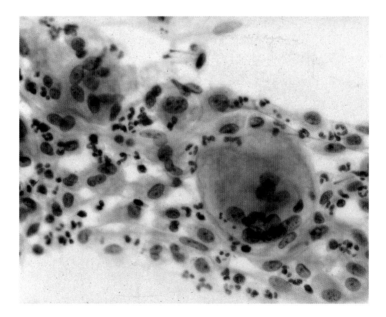

FIGURE 2.30. Atrophy with multinucleated giant cells (CP). Multinucleated histiocytic giant cells are a nonspecific finding and are often seen in postmenopausal and postpartum specimens.

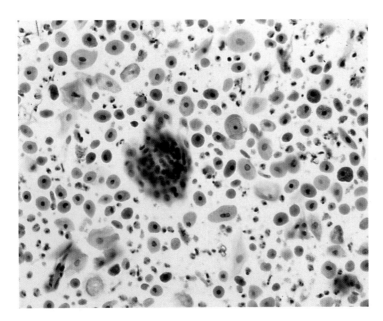

[*] FIGURE 2.31. Atrophy (LBP). Note more dissociation of parabasal cells and degenerated parabasal cells in a relatively clean background.

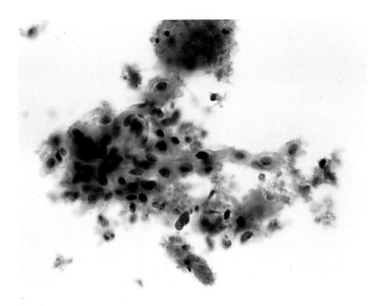

[*] FIGURE 2.32. Atrophy with inflammation (atrophic vaginitis) (LBP). In LBP, the granular debris is clumped and adheres to cell clusters in a pattern that may mimic "clinging tumor diathesis" (see Fig. 5.39). Attention to cellular features is crucial.

Other Non-Neoplastic Findings Not Specifically Listed in 2001 Bethesda Terminology

Tubal Metaplasia (Figs. 2.33–2.35)

Criteria

Columnar endocervical cells that may occur in small groups or pseudo-stratified, often crowded groups (Fig. 2.34).

Nuclei are round to oval and may be enlarged, pleomorphic, and often hyperchromatic.

Chromatin is evenly distributed and nucleoli are usually not seen.

Nuclear to cytoplasmic ratio can be high.

The cytoplasm may show discrete vacuoles or goblet cell change (Fig. 2.35).

Presence of cilia and/or terminal bars is characteristic, but single ciliated cells in isolation are not sufficient for the designation.

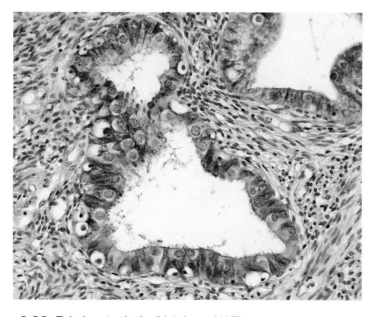

FIGURE 2.33. Tubal metaplasia (histology, H&E).

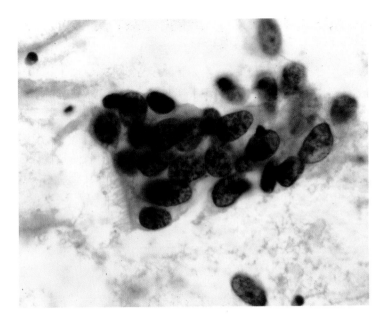

FIGURE 2.34. Tubal metaplasia (CP). Note terminal bar and cilia at left edge.

[*] **FIGURE 2.35.** Tubal metaplasia (CP). There are ciliated columnar endo-cervical cells, and a goblet cell is also seen in the center.

Keratotic Cellular Changes ("Typical Parakeratosis") (Figs. 2.36, 2.37)

Criteria

Miniature superficial squamous cells with dense orangeophilic or eosinophilic cytoplasm.

Cells may be seen in isolation, in sheets, or in whorls; cell shape may be round, oval, polygonal, or spindle shaped.

Nuclei are small and dense (pyknotic).

If atypical nuclear or cellular changes are present, consider atypical squamous cell (ASC) interpretation.

Keratotic Cellular Changes ("Hyperkeratosis") (Fig. 2.38)

Criteria

Anucleate but otherwise unremarkable mature polygonal squamous cells, often associated with mature squamous cells with keratohyaline granules.

Empty spaces or "ghost nuclei" may be seen.

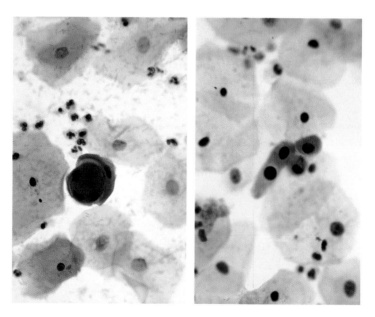

[*] FIGURE 2.36. Keratotic cellular changes, "typical parakeratosis" (CP). On the *left side*, note the "squamous pearl" formation in this specimen from a 49-year-old woman being followed up after treatment for SIL. On the *right side* is a small cluster of miniature squamous cells. Both are examples of "typical parakeratosis" with small bland nuclei.

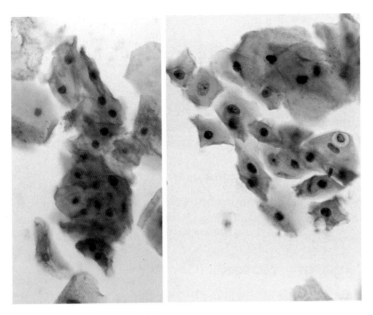

[★] **Figure 2.37.** Keratotic cellular changes, "typical parakeratosis" (*left*, CP; *right*, LBP). On the left is an orangeophilic cluster, and on the right are more eosinophilic squamous cells with small, opaque nuclei. Human papillomavirus (HPV) testing, performed for other reasons on the liquid-based specimen, was negative.

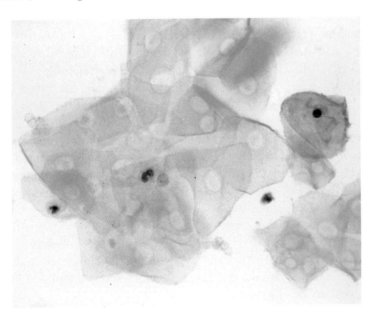

[★] **Figure 2.38.** Keratotic cellular changes, "hyperkeratosis" (LBP). Anucleate, mature polygonal squamous cells with ghostlike "nuclear holes." (Reprinted with permission from Williamson, BA, DeFrias, D., Gunn, R., Tarjan, G., Nayar, R. Significance of extensive hyperkeratosis on cervical vaginal smears. *Acta Cytol* 2003;47(5):750.)

Lymphocytic (Follicular) Cervicitis (Figs. 2.39, 2.40)

Criteria

Polymorphous populations of lymphocytes with or without tingible body macrophages are seen in clusters or streaming out in mucus (Fig. 2.39).

Liquid-Based Preparations

Lymphoid cells may appear in clusters and more scattered single cells can be seen in the background (Fig. 2.40).

Explanatory Notes

Reactive changes are included in the 2001 Bethesda System under NILM. Recognizing such changes is important for defining the boundary between NILM and epithelial abnormalities. In general, round nuclear contours and even chromatin distribution favor a non-neoplastic process. However, exuberant reactive changes in mature squamous cells may raise the differential of "low-grade squamous intraepithelial lesion" (LSIL) or even cancer if reparative features are present (see following paragraph on repair), and an interpretation of "atypical squamous cells of undetermined significance" (ASC-US) or "atypical squamous cells, cannot exclude high grade SIL" (ASC-H) may be considered. Reactive nuclear enlargement in squamous metaplastic cells may mimic "high-grade SIL" (HSIL).

Squamous metaplastic cells in LBPs, in particular, often demonstrate an increased nuclear/cytoplasmic (N:C) ratio due to rounding up of cells, which may raise the differential of an HSIL. In addition, overlapping nuclei in a binucleated cell may give the impression of hyperchromasia (see Fig. 4.21). An N:C ratio of less than 50%, smooth nuclear contours, and even distribution of chromatin all favor benign squamous metaplasia. A higher N:C ratio in conjunction with hyperchromasia and/or nuclear contour irregularities such as notching or grooving should prompt consideration of HSIL or an ASC-H designation.[19] Note that with degeneration, nuclei may become wrinkled and hyperchromatic and therefore difficult to differentiate from HSIL; in such cases, an interpretation of ASC-H may be appropriate (see Fig. 4.12).

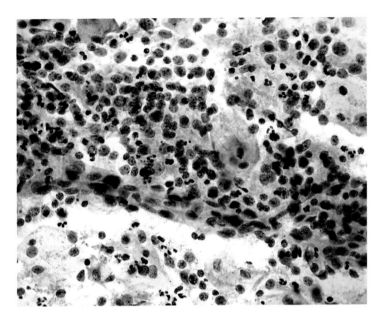

FIGURE 2.39. Lymphocytic (follicular) cervicitis (CP). Abundant lymphoid cells with a tingible body macrophage located centrally.

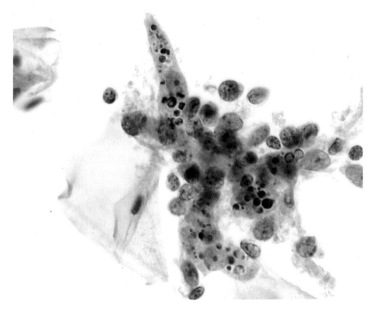

FIGURE 2.40. Lymphocytic (follicular) cervicitis (LBP). Note polymorphous population of lymphoid cells and tingible body macrophages; cells may clump on LBPs.

Reparative changes ("typical repair") may involve mature squamous, squamous metaplastic, or columnar epithelium. The increased nuclear size and prominent nucleoli characteristic of repair may raise concern for the presence of a more significant lesion. However, in a reparative process, cells typically occur in monolayer sheets with nuclei oriented in the same direction, imparting a streaming look to the epithelial fragments. In addition, there is a paucity of single cells with nuclear changes. If marked anisonucleosis, irregularities in chromatin distribution, or variation in size and shape of nucleoli are present, so-called "atypical repair," the changes should be categorized as "atypical glandular cells" or "atypical squamous cells."

Acute radiation-induced changes, consisting of degenerated blood, bizarre cell forms, and cellular debris, generally resolve within six months following therapy. However, in some patients, chronic radiation-induced cellular changes may persist indefinitely. These chronic changes can include cytomegaly, nuclear enlargement without nuclear/cytoplasmic ratio alteration, mild hyperchromasia, engulfed neutrophils, and persistent polychromatic cytoplasmic staining. Certain chemotherapeutic agents may produce changes in cervical epithelial cells similar to those seen with acute and chronic radiation effects.

The reactive glandular cell clusters occasionally seen in women with IUDs may represent either endometrial or endocervical columnar cells exfoliated as a result of chronic irritation by the device. Such cells may persist for several months after removal of the IUD. Cells may be shed in two patterns: as three-dimensional clusters or singly. The three-dimensional glandular clusters with vacuolated cytoplasm and nuclear changes may closely resemble clusters of cells derived from adenocarcinoma of the endometrium, fallopian tube, or ovary (see Fig. 6.5). Single cells with higher nuclear/cytoplasmic ratio may mimic a high-grade SIL; however, the morphologic spectrum of abnormalities usually present with true precursor lesions is absent. In general, the diagnosis of adenocarcinoma should be made only with caution in the presence of an IUD. If there is any doubt as to the significance of the cellular abnormalities, the cytopathologist should consider recommending removal of the IUD followed by repeat cervical cytology sampling.

On occasion, benign-appearing glandular cells may be seen posthysterectomy.[20] There are a number of explanations for this phenomenon, including development of adenosis after traumatic stimulation of stromal mesenchymal cells,[21,22] mucinous or goblet cell metaplasia in response to atrophy,[23] or prolapse of the remaining fallopian tube after simple hysterectomy. Most important is to exclude adenocarcinoma, particularly when the hysterectomy was performed for glandular neoplasia.

Atrophy is a normal aging phenomenon with a wide spectrum of cellular changes and variable amounts of inflammation. Note that air-drying, a common problem with smear preparations of this type of specimen, may artificially cause nuclear enlargement. Reporting of atrophic changes is variable and nonreproducible.[24] However, atypical cellular changes associated with atrophy or atrophy with inflammation may warrant an interpretation of atypical squamous cells (ASC) (see Figs. 4.17, 4.18).

Multinucleated histiocytes are frequently observed in cervical cytology specimens and can be associated with chronic inflammatory processes and as part of granulation tissue along with occasional fibroblasts and poorly preserved epithelial cells. They may be numerous in specimens from postmenopausal women in the absence of inflammation, and can also be seen postpartum.[25]

Tubal metaplasia of endocervical cells is also a normal phenomenon that includes several cell types (ciliated cells, peg cells, goblet cells).[26] It is important to recognize that enlarged and/or crowded nuclei and nuclear stratification may lead to an interpretation of atypical endocervical glandular cells unless terminal bars and cilia are identified (see Figs. 6.13–6.16).

Normally, the cervix is a nonkeratinizing, stratified squamous epithelium. Keratotic changes may occur as a reactive phenomenon or in association with human papillomavirus (HPV)-induced cell changes. "Keratosis," "hyperkeratosis," "parakeratosis," and "dyskeratosis" are descriptive terms for such changes, but have been used inconsistently in the past, sometimes to convey benignity but in other cases to indicate a dysplastic change. These terms are not specifically listed in Bethesda terminology due to lack of consensus definitions; they are included parenthetically for clarification only. Although some cytologists may choose to include such terms to describe a morphologic feature, they do not constitute a report interpretation and should not be used alone to report results.

The Bethesda classification and interpretation of such keratotic changes depend on the cytoplasmic and nuclear alterations present. Miniature squamous cells with small pyknotic nuclei and orangeophilic to eosinophilic cytoplasm ("parakeratosis") is a non-neoplastic reactive cellular change. However, single cells or cell clusters that demonstrate pleomorphism of cell size or shape, that is, caudate or elongate cells, and/or increased nuclear size and chromasia ("atypical parakeratosis" or "dyskeratosis"), represent an epithelial cell abnormality. Such findings should be categorized as atypical squamous cells (ASC) or as a squamous intraepithelial lesion (SIL), depending on the degree of cellular abnormality identified (see Figs. 4.3, 4.17, 5.8, 5.23, 5.24).

Anucleate, but otherwise unremarkable mature, squamous cells ("hyperkeratosis") is a non-neoplastic change. Inadvertent contamination of the specimen with vulvar material may also introduce anucleate squamous cells on the cervical cytology specimen. When extensive hyperkeratosis is seen an underlying neoplastic or nonneoplastic process may be present.[27] Thick plaques of pleomorphic anucleate squamous cells with irregular contours may rarely be the only clue to an underlying squamous cell carcinoma.[28]

Lymphocytic cervicitis (follicular cervicitis) is an uncommon form of chronic cervicitis that results in the formation of mature lymphoid follicles in the subepithelium of the uterine cervix.

Sample Reports

Example 1:
Specimen Adequacy:
Satisfactory for evaluation; endocervical/transformation zone component present.
Interpretation:
Negative for intraepithelial lesion or malignancy.

Example 2:
Specimen Adequacy:
Satisfactory for evaluation; endocervical/transformation zone component present; partially obscuring inflammation present.
Interpretation:
Negative for intraepithelial lesion or malignancy.
Trichomonas vaginalis identified.
Reactive squamous cells associated with inflammation.

Example 3:
Specimen Adequacy:
Satisfactory for evaluation; endocervical/transformation zone component absent.
Interpretation:
Negative for intraepithelial lesion or malignancy.
Reactive cellular changes associated with radiation.

Example 4:
Specimen Adequacy:
Satisfactory for evaluation; endocervical/transformation zone component cannot be assessed because of severe atrophy.
Interpretation:
Negative for intraepithelial lesion or malignancy.
Fungal organisms morphologically consistent with *Candida* species.

Atrophy.

Bethesda System 2001 Workshop Forum Group Moderators:

Nancy A. Young, M.D., Marluce Bibbo, M.D., Sally-Beth Buckner, S.C.T. (ASCP), Terence Colgan, M.D., Dorothy Rosenthal, M.D., Edward Wilkinson, M.D.

References

1. Colgan TJ, Woodhouse SL, Styer PE, et al. Reparative changes and the false-positive/false-negative Papanicolaou test. *Arch Pathol Lab Med* 2001;125(1):134–140.
2. Young NA, Naryshkin S, Atkinson BF, et al. Interobserver variability of cervical smears with squamous-cell abnormalities: a Philadelphia study. *Diagn Cytopathol* 1994;11(4):352–357.
3. Young NA, Kline TS. Benign cellular changes: allied ambiguity in CLIA '88 and the Bethesda System [editorial]. *Diagn Cytopathol* 1994;10(4):307–308.
4. Interlaboratory Comparison Program in Cervicovaginal Cytopathology (PAP). 1992 PAP. Supplementary Questionnaire.
5. Davey DD, Nielsen ML, Frable WJ, et al. Improving accuracy in gynecologic cytology. Results of the College of American Pathologists Interlaboratory Comparison Program in Cervicovaginal Cytology [see comments]. *Arch Pathol Lab Med* 1993; 117(12):1193–1198.
6. Young NA. Back to the negative Pap test: behind the scenes at Bethesda 2001. *Diagn Cytopathol* 2002;26(4):207–208.
7. Jones BA. Rescreening in gynecologic cytology. Rescreening of 3762 previous cases for current high-grade squamous intraepithelial lesions and carcinoma: a College of American Pathologists Q-Probes study of 312 institutions. *Arch Pathol Lab Med* 1995;119(12):1097–1103.
8. Barr Soofer S, Sidawy MK. Reactive cellular change: is there an increased risk for squamous intraepithelial lesions? [see comments]. *Cancer* (Phila) 1997;81(3):144–147.
9. Malik SN, Wilkinson EJ, Drew PA, et al. Benign cellular changes in Pap smears. Causes and significance. *Acta Cytol* 2001;45(1):5–8.
10. Giacomini G, Paavonen J, Rilke F. Microbiologic classification of cervicovaginal flora in Papanicolaou smears. *Acta Cytol* 1989;33(2):276–278.
11. Giacomini G, Schnadig VJ. The cervical Papanicolaou smear: bacterial infection and the Bethesda System. *Acta Cytol* 1992;36(1):109–110.
12. Bartlett JG, Moon NE, Goldstein PR, et al. Cervical and vaginal bacterial flora: ecologic niches in the female lower genital tract. *Am J Obstet Gynecol* 1978;130(6):658–661.
13. Prey M. Routine Pap smears for the diagnosis of bacterial vaginosis. *Diagn Cytopathol* 1999;21(1):10–13.
14. Donders GG, Van Bulck B, Caudron J, et al. Relationship of bacterial vaginosis and

mycoplasmas to the risk of spontaneous abortion. *Am J Obstet Gynecol* 2000;183(2): 431–437.

15. Schwebke JR. Bacterial vaginosis. *Curr Infect Dis Rep* 2000;2(1):14–17.

16. Fiorino AS. Intrauterine contraceptive device-associated actinomycotic abscess and actinomyces detection on cervical smear. *Obstet Gynecol* 1996;87(1):142–149.

17. Banuelos Panuco CA, Deleon Rodriguez I, Hernandez Mendez JT, et al. Detection of *Chlamydia trachomatis* in pregnant women by the Papanicolaou technique, enzyme immunoassay and polymerase chain reaction. *Acta Cytol* 2000;44(2):114–123.

18. Edelman M, Fox A, Alderman E, et al. Cervical Papanicolaou smear abnormalities and *Chlamydia trachomatis* in sexually active adolescent females. *J Pediatr Adolesc Gynecol* 2000;13(2):65–69.

19. Sherman ME, Solomon D, Schiffman M. Qualification of ASCUS: a comparison of equivocal LSIL and equivocal HSIL cervical cytology in the ASCUS LSIL Triage Study. *Am J Clin Pathol* 2001;116:386–394.

20. Ponder TB, Easley KO, Davila RM. Glandular cells in vaginal smears from posthysterectomy patients. *Acta Cytol* 1997;41:1701–1704.

21. Gondos B, Smith LR, Townsend DE. Cytologic changes in cervical epithelium following cryosurgery. *Acta Cytol* 1970;14(7):386–389.

22. Sedlacek TV, Riva JM, Magen AB, et al. Vaginal and vulvar adenosis. An unsuspected side effect of CO_2 laser vaporization. *J Reprod Med* 1990;35(11):995–1001.

23. Bewtra C. Columnar cells in posthysterectomy vaginal smears. *Diagn Cytopathol* 1992;8(4):342–345.

24. *2001 Interlaboratory Comparison Program in Cervicovaginal Cytopathology (PAP) Year End Summary Report.* Northfield: College of American Pathologists, 2002.

25. Koss LG. Inflammatory processes and other benign disorders of the cervix and vagina. In: *Diagnostic Cytology and Its Histopathologic Bases.* Philadelphia: Lippincott, 1992:314–370.

26. Babkowski RC, Wilbur DC, Rutkowski MA, et al. The effects of endocervical canal topography, tubal metaplasia, and high canal sampling on the cytologic presentation of non-neoplastic endocervical cells. *Am J Clin Pathol* 1996;105:403–410.

27. Williamson BA, DeFrias DVS, Gunn R, et al. Significance of extensive hyperkeratosis on cervical/vaginal smears. *Acta Cytol* 2003;47:749–752.

28. Bibbo M, Wied GL. Look-alikes in gynecologic cytology. In: Wied GL, ed. *Tutorials of Cytology*, Vol. 12, 2nd Ed. Chicago: Tutorials of Cytology Press, 1988:3.

Chapter 3

Endometrial Cells: The How and When of Reporting

Ann T. Moriarty and Edmund S. Cibas

Other

❏ Endometrial cells (*in a woman ≥40 years of age*)
  (*Specify if "negative for squamous intraepithelial lesion"*)

Background

Exfoliated endometrial cells are commonly seen in specimens obtained during the proliferative phase of the menstrual cycle. However, endometrial cells have been considered a potential harbinger of endometrial adenocarcinoma when seen in cervical/vaginal cytology preparations of postmenopausal women or outside of the proliferative phase of the menstrual cycle. In the 1991 Bethesda System, "cytologically benign appearing" endometrial cells in postmenopausal women were reported as an "Epithelial cell abnormality." This perspective was based upon retrospective reviews performed in the 1970s on the significance of endometrial cells detected in cervical smears. In the largest of these early studies, follow-up showed that women over 40 years of age occasionally demonstrated endometrial abnormalities, whereas women under 40 did not have endometrial pathology.[1,2] A more recent study suggests that most women with endometrial carcinoma present with bleeding symptoms,[3] whereas other studies note that the presence of endometrial cells was the only abnormal finding in a small proportion of asymptomatic women in whom endometrial adenocarcinoma was detected.[4,5]

An individual woman's risk factors for endometrial carcinoma, clinical symptoms, menstrual history, hormone therapy, and menopausal status are often unclear, inaccurate, or unknown to the laboratory. Therefore, in the 2001 Bethesda System, the presence of exfoliated endometrial cells is reported in all women 40 or older; the general categorization, if used, is "Other." *Atypical* endometrial cells should still be reported under the general category of "Epithelial cell abnormality."

It should also be emphasized that cervical cytology is primarily a screening test for squamous intraepithelial lesions and squamous cell carcinoma. It is unreliable for the detection of endometrial lesions and should not be used to evaluate causes of suspected endometrial abnormalities.

Exfoliated Endometrial Cells (Figs. 3.1–3.4)

Criteria

Exfoliated cells occur in ball-like clusters and rarely as single cells (Figs. 3.1, 3.2).

Nuclei are small, round, and approximate the area of a normal intermediate cell nucleus.

Nuclear chromatin pattern is often difficult to discern because of three-dimensional rounding of the cell clusters.

Nucleoli are inconspicuous.

Cytoplasm is scant, basophilic, and occasionally vacuolated.

Cell borders are ill defined.

During the first half of the menstrual cycle, double-contoured clusters of endometrial cells ("exodus" pattern) may be seen (Fig. 3.3).

Liquid-Based Preparations

Cell groups may appear "above the plane" of squamous epithelial cells, especially in gradient-based methods.

Cells appear in tight three-dimensional clusters, loose clusters, and singly.

Single cells may be more evident in the cleaner background.

Nuclei may be bean shaped; nucleoli and chromatin detail may be more apparent (Fig. 3.2).

Intracytoplasmic vacuoles are more common and easily visible.

Single cell necrosis (apoptosis) is easily seen in the endometrial cell groups of liquid-based preparations (Fig. 3.4).

Background appears cleaner, especially in menstrual smears (Fig, 3.3), compared to conventional preparations.

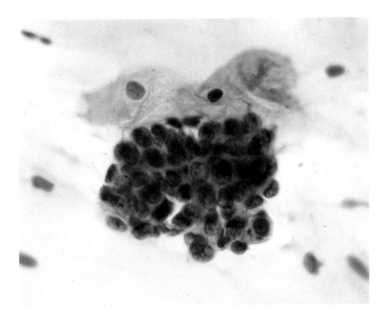

FIGURE 3.1. Exfoliated endometrial cells (conventional preparation, CP). 45-year-old woman, menstrual specimen. Three-dimensional clusters of endometrial cells with small, round nuclei similar in size to an intermediate squamous cell nucleus. Nucleoli are inconspicuous. Cytoplasm is scant, and cell borders are indistinct.

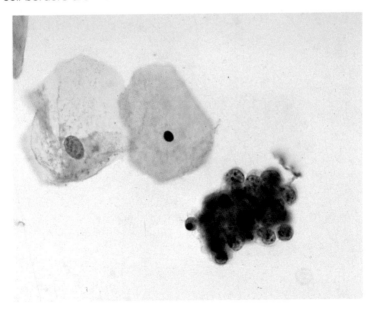

[*] FIGURE 3.2. Exfoliated endometrial cells (liquid-based preparation, LBP). 36-year-old woman, day 1 of menstrual cycle. Three-dimensional cluster of endometrial cells. Nuclei may be bean shaped with more apparent chromocenters in LBPs than CPs. Follow up was NILM.

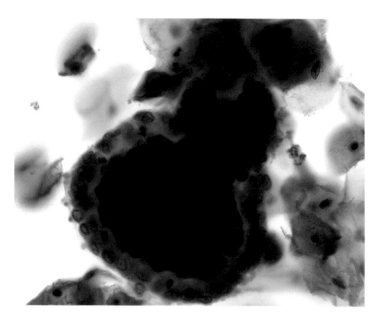

[*] FIGURE 3.3. Exodus pattern of exfoliated endometrial cells (LBP). 43-year-old woman. There is a double contoured pattern with glandular epithelial cells surrounding a dark core of stromal cells. Note the cleaner background typically seen in LBP menstrual specimens. Follow up was NILM. [[*] Bethesda Interobserver Reproducibility Project (BIRP) image (see xviii Introduction).]

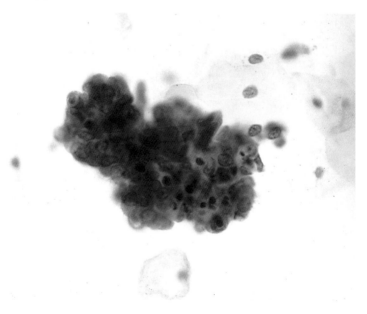

FIGURE 3.4. Exfoliated endometrial cells (LBP). 42-year-old woman in first half of menstrual cycle. Single cell necrosis (apoptosis) can be seen in exfoliated endometrial groups in LBPs.

Explanatory Notes

In the 2001 Bethesda System (TBS), only exfoliated, intact endometrial cells clusters should be reported in women 40 years or older. Exfoliated groups of endometrial cells may be of epithelial and/or stromal origin; however, morphologic distinction of these two cell types is not usually possible. In liquid-based preparations, exfoliated endometrial cells may be slightly larger with more obvious nucleoli and enhanced chromatin detail, compared to conventional smears. These features may be worrisome to those unfamiliar with the appearance of endometrial cells in liquid-based preparations.

Abraded endometrial cells or lower uterine segment (LUS) fragments, abraded stromal cells, and histiocytes, when seen in the absence of exfoliated endometrial cells, are not associated with increased risk of endometrial cancer and therefore do not generally warrant reporting.[6,7,8] LUS may be seen as a result of vigorous endocervical sampling and is characteristically composed of biphasic tissue fragments with densely packed spindle cells, vascular spaces, and simple or branching tubular glands embedded within the stroma[9] (Fig. 3.5). Glandular and stromal cells from inadvertently sampled endometrium during the proliferative

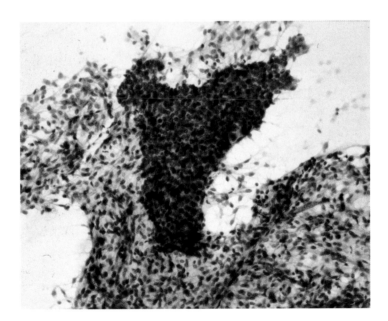

[*] FIGURE 3.5. Lower uterine segment (LUS) fragments (CP). Routine screen from 45-year-old woman. Epithelial fragments are embedded in vascular stroma composed of tightly packed spindled cells. LUS represents abraded endometrium and does not carry the same implications as exfoliated cells if seen in a woman ≥ 40. Follow up was NILM for 5 years.

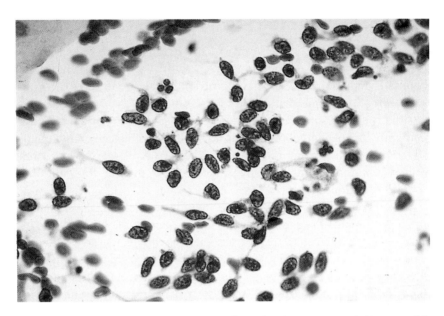

FIGURE 3.6. Endometrial stromal cells (CP). Routine screen of 43-year-old woman. "Deep" stromal cells are spindle shaped with scant cytoplasm and round to oval nuclei. Stromal cells do not indicate increased risk for the presence of endometrial carcinoma.

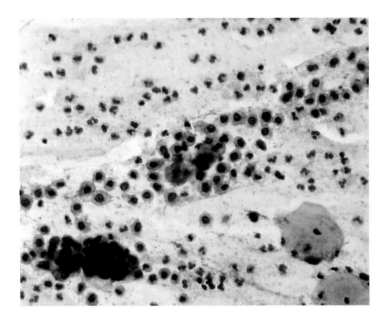

FIGURE 3.7. Histiocytes (CP). Routine screen of 30-year-old woman, day 7 of menstrual cycle. Histiocytes have reniform to round nuclei, with moderate, finely vacuolated cytoplasm. They are often seen in association with exodus in the first half of the menstrual cycle. Histiocytes alone are not considered significant in predicting the presence of endometrial carcinoma.

phase may have abundant mitoses. Abraded "deep stromal" endometrial cells may vary from round to spindle shaped and have small, oval nuclei and scant cytoplasm (Fig. 3.6). Histiocytes are often associated with, and may be indistinguishable from, superficial stromal cells.[6,8] Both demonstrate single cells with moderate, vacuolated cytoplasm and bean-shaped to round nuclei (Fig. 3.7).

Loose clusters of naked nuclei showing smooth nuclear contours, molding, and evenly distributed granular chromatin have been reported in women receiving tamoxifen therapy and also in atrophic smears. It is postulated that the incidence of such "small cells" increases with age regardless of tamoxifen therapy, but that the cells/naked nuclei are more easily detected in the background of estrogenized squamous cells due to tamoxifen.[10,11] The naked nuclei may be of parabasal or reserve cell origin and should not be mistaken for endometrial cells (Fig. 3.8).

An optional educational comment is recommended for use when reporting exfoliated endometrial cells in a woman who is 40 years or older. Any comment should stress that exfoliated endometrial cells are usually derived from a benign process and that only a small proportion of women

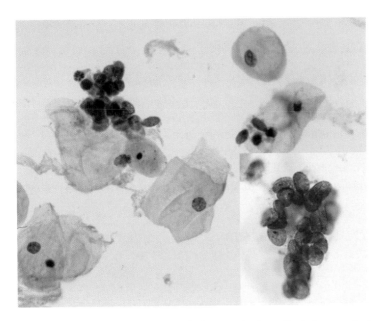

FIGURE 3.8. Naked nuclei seen here in loose clusters that demonstrate molding and granular chromatin. The *insert (lower right)* shows a higher magnification of a grapelike cluster with bland chromatin (LBP). This postmenopausal patient was receiving tamoxifen therapy. Such clusters should not be mistaken for endometrial cells.

with this finding have endometrial abnormalities. If the date of the last menstrual period (LMP) is provided, and the specimen was obtained in the first half of the cycle, the laboratory may wish to append a comment indicating that the finding of endometrial cells correlates with the menstrual history (see sample report 3B).

It is up to the laboratory to have a policy specifying the circumstances under which endometrial cells—that are not cytologically atypical—are referred for pathologist's review.

Sample Reports

1. Using a General Categorization:
   General Categorization:
   Other: See Interpretation/Result.
   Interpretation/Result:
   Endometrial cells present in a woman ≥40 years of age (see note).
   Negative for squamous intraepithelial lesion.

2. Without Use of the Optional General Categorization: Other
   Endometrial cells are present in a woman ≥40 years of age (see note).
   Negative for squamous intraepithelial lesion.

3. Educational Note(s) (optional):
   A. For all reports with endometrial cells in women 40 years or older:
      Endometrial cells after age 40, particularly out of phase or after menopause may be associated with benign endometrium, hormonal alterations and less commonly, endometrial/uterine abnormalities. Clinical correlation is recommended.
   B. Additional note to consider when a woman's LMP is provided and endometrial cells are seen in the first half of the menstrual cycle:
      Endometrial cells correlate with the menstrual history provided.

Bethesda System 2001 Workshop Forum Group Moderators:

Edmund Cibas, M.D., Gary W. Gill, C.T. (ASCP), Meg McLachlin, M.D., Ann T. Moriarty, M.D., Ellen Sheets, M.D., Theresa M. Somrak, J.D., C.T. (ASCP), Rosemary E. Zuna, M.D.

References

1. Ng ABP, Reagan JW, Hawliczek S, et al. Significance of endometrial cells in the detection of endometrial adenocarcinoma and its precursors. *Acta Cytol* 1974;18:356–361.
2. Gondos B, King EB. Significance of endometrial cells in cervicovaginal smears. *Ann Clin Lab Sci* 1977;7:486–490.
3. Gomez-Fernandez CR, Ganjei-Azar P, Capote-Dishaw J, et al. Reporting normal endometrial cells in Pap smear: an outcome appraisal. *Gynecol Oncol* 1999;74:381–384.
4. Zucker PK, Kadson EJ, Feldstein ML. The validity of Pap smear parameters as predictors of pathology in menopausal women. *Cancer* 1985;56:2256–2263.
5. Cherkis RC, Patten SF, Andrews TJ, et al. Significance of normal endometrial cells detected by cervical cytology. *Obstet Gynecol* 1998;72:242–244.
6. Ngyuen TN, Bourdeau J-L, Ferenczy A, et al. Clinical significance of histiocytes in the detection of endometrial adenocarcinoma and hyperplasia. *Diagn Cytopathol* 1998;19:89–93.
7. Chang A, Sandweiss L, Bose S. Cytologically benign endometrial cells in the Papanicolaou smears of postmenopausal women. *Gynecol Oncol* 2001;80:37–43.
8. Nassar A, Fleisher SR, Nasuti JF. Value of histiocyte detection in Pap smears for predicting endometrial pathology. An institutional experience. *Acta Cytol* 2003;47:762–767.
9. De Peralta-Venturino MN, Purslow MJ, Kini SR. Endometrial cells of the "lower uterine segment" (LUS) in cervical smears obtained by endocervical brushings: a source of potential diagnostic pitfall. *Diagn Cytopathol* 1995;12:263–271.
10. Opjorden SL, Caudill JL, Humphrey SK, et al. Small cells in cervical-vaginal smears of patients treated with tamoxifen. *Cancer* 2001;93(1):23–28.
11. Yang YJ, Trapkin LK, Demoski RK, et al. The small blue cell dilemma associated with tamoxifen therapy. *Arch Pathol Lab Med* 2001;125:1047–1050.

Chapter 4

Atypical Squamous Cells

Mark E. Sherman, Fadi W. Abdul-Karim, Jonathan S. Berek,
Celeste N. Powers, Mary K. Sidawy, and Sana O. Tabbara

Epithelial Cell Abnormalities

Squamous cell

- ❏ Atypical squamous cells (ASC)
  - ➤ of undetermined significance (ASC-US)
  - ➤ cannot exclude HSIL (ASC-H)

Background

The forerunner of the new category "Atypical squamous cells" (ASC)[1] was the more broadly defined interpretation of atypical squamous cells of undetermined significance.[2] A national survey of reporting practices in 768 laboratories during 1996 found that the classification of atypical squamous cells of undetermined significance accounted for a mean of 5.2% and a median of 4.5% of all cervical cytology reports.[3] These data indicate that many laboratories in the United States were unable to meet the target of maintaining a reporting frequency for atypical squamous cells of undetermined significance under 5%.[4] These findings prompted the development of the new category, ASC, which has a narrower definition and a simple dichotomous system of qualifiers.

ASC does not represent a single biologic entity; it subsumes changes that are unrelated to oncogenic human papillomavirus (HPV) infection and neoplasia as well as findings that suggest the possible presence of underlying cervical intraepithelial neoplasia (CIN) and rarely carcinoma. In screening programs representative of the U.S. population, approximately 50% of women with ASC are infected with high-risk/oncogenic types of human papillomaviruses (HPV).[5] The remaining noninfected women are not at increased cancer risk. Numerous non-neoplastic conditions that are unrelated to HPV infections may produce cytologic mimics classified as

ASC, including inflammation, air-drying, atrophy with degeneration, and other artifacts. In many instances, the process that resulted in the ASC interpretation remains undefined, even following diagnostic workup.

Data demonstrate that simply eliminating the ASC category by classifying every specimen as "negative for intraepithelial lesion/malignancy" (NILM) or "squamous intraepithelial leson" (SIL) using light microscopy alone is not feasible; misclassification may lead to loss of both sensitivity and positive predictive value.[6] Accordingly, the 2001 Bethesda System maintained an equivocal category (ASC) and simplified its qualifiers to realistically reflect the inability of pathologists to accurately and reproducibly interpret these specimens. The two most common qualifiers of atypical squamous cells in the 1991 Bethesda classification, "Favor reactive," and "Not otherwise specified," have been eliminated in the new Bethesda System.[1] All interpretations of ASC should be qualified as "Of undetermined significance" (ASC-US) or "cannot exclude HSIL" (ASC-H).

ASC-US refers to changes that are either suggestive of LSIL or SIL of indeterminate grade. Although most ASC-US interpretations are suggestive of LSIL, the qualifier "undetermined significance" is preferred because approximately 10% to 20% of women with ASC-US prove to have an underlying CIN 2 or CIN 3.[5] ASC-US is expected to comprise more than 90% of ASC interpretations in most laboratories. ASC-H is a designation reserved for the minority of ASC cases (expected to represent less than 10%) in which the cytologic changes are suggestive of HSIL. Only equivocal specimens specifically worrisome for HSIL should be distinguished from the bulk of ASC using the designation ASC-H. Cases classified as ASC-H are associated with a higher positive predictive value for detecting an underlying CIN 2 or CIN 3 than ASC-US, but are less predictive of CIN 2 or worse than definitive interpretations of HSIL.[7,8]

Atypical Squamous Cells

Definition

ASC refers to cytologic changes *suggestive* of SIL, which are qualitatively or quantitatively insufficient for a definitive interpretation.[1] Cytologic findings that are suggestive of benign reactive changes should be carefully reviewed and judiciously classified as "negative for intraepithelial lesion or malignancy" whenever possible.

The interpretation of ASC requires that the cells in question demonstrate three essential features: (1) squamous differentiation, (2) increased ratio of nuclear to cytoplasmic area, and (3) minimal nuclear hyperchromasia, chromatin clumping, irregularity, smudging, or multinucleation. Unequivocally normal-appearing cells on the same slide should be used for comparison in determining whether the interpretation of ASC is warranted (see below). Abnormal-appearing nuclei are a prerequisite for the interpretation of ASC; however, cytoplasmic changes that are often associated with HPV infection such as dense orangeophilia (parakeratosis) and perinuclear halos (koilocytosis) should prompt a careful search for cells that warrant an interpretation of ASC or SIL.

The ASC category was developed to designate the interpretation of an entire specimen, not individual cells. This consideration combined with the subtle and subjective findings in these specimens have resulted in poor reproducibility, compounding the difficulty in developing and illustrating strict criteria.[9] Furthermore, the infinite appearances that ASC may assume, including nonphotogenic degenerative and artifactual changes, permit only a fractional representation of changes that experts might accept if not agree upon as ASC.

Atypical Squamous Cells of Undetermined Significance (ASC-US) (Figs. 4.1—4.7)

Criteria

Nuclei are approximately two and one half to three times the area of the nucleus of a normal intermediate squamous cell (approximately 35 μm^2).[10]

Slightly increased ratio of nuclear to cytoplasmic area (N/C).

Minimal nuclear hyperchromasia and irregularity in chromatin distribution or nuclear shape.

Nuclear abnormalities associated with dense orangeophilic cytoplasm ("atypical parakeratosis") (Fig. 4.3).

Liquid-Based Preparations

The appearance of ASC-US in smears and liquid-based cytology is similar; in smears, cells may appear larger and flatter (Figs. 4.1, 4.2, 4.7).

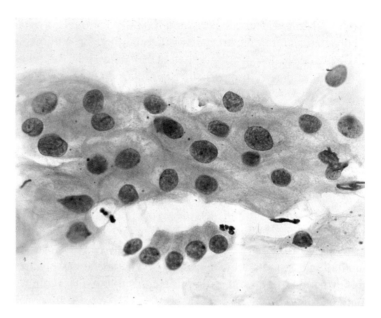

[*] FIGURE 4.1. Negative for intraepithelial lesion or malignancy (NILM) versus Atypical squamous cells of undetermined significance (ASC-US) (conventional preparation, CP). Premenopausal woman. Mature squamous cells show mild nuclear enlargement, binucleation, and even chromatin distribution. Note benign endocervical cells at bottom of field. [[*] Bethesda Interobserver Reproducibility Project (BIRP) image (see xviii Introduction).]

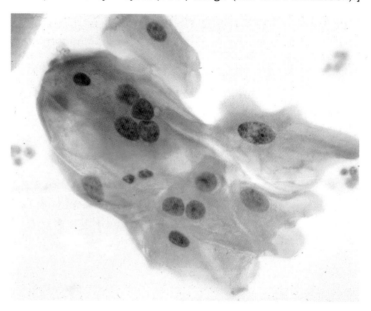

FIGURE 4.2. ASC-US (CP). Cells with multinucleation, nuclear enlargement, and air-drying artifact, possibly representing LSIL.

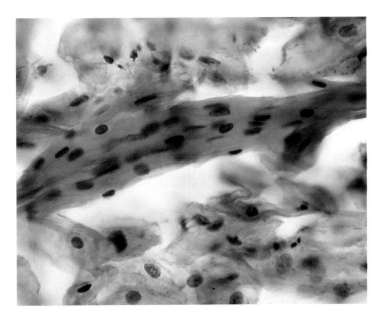

[*] **Figure 4.3.** ASC-US (CP). Plaque of cells with dense orangeophilic cytoplasm and minimally irregular, hyperchromatic nuclei.

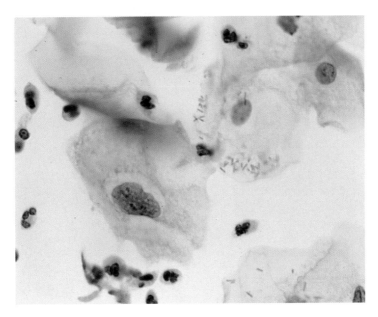

[*] **Figure 4.4.** ASC-US (liquid-based preparation, LBP). Routine screen 32-year-old woman. Single multinucleated cell with cytoplasmic halo in a background of inflammation. Adjacent squamous cell shows adherent lactobacilli.

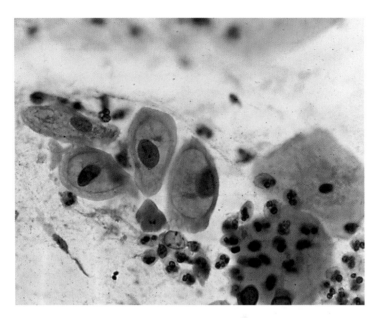

[*] Figure 4.5. ASC-US (CP). Premenopausal woman. Cells with central cytoplasmic clearing; findings suggest either LSIL (HPV effect) or glycogenation. (Reprinted with permission from Kurman, RJ (ed.), *Blaustein's Pathology of the Female Genital Tract*, Fourth Edition, Springer-Verlag New York, © 1994.)

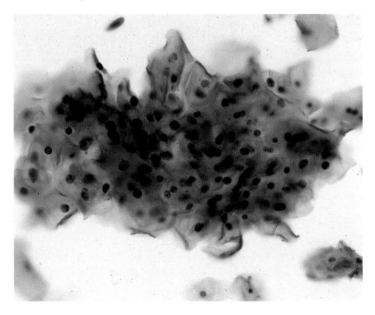

[*] Figure 4.6. ASC-US (LBP). 21-year-old woman. Thick cohesive sheet of cells with focal nuclear enlargement, orangeophilic cytoplasm, and binucleation. Follow up was CIN 1.

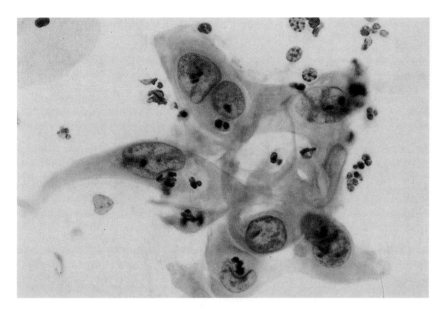

FIGURE 4.7. ASC-US (CP). Group of cells with features of repair; however, the presence of irregular chromatin distribution and the increased N/C ratio are not typical (see Figs. 2.19, 2.20). Atypical reparative squamous cells may be classified as ASC-US or sometimes as ASC-H.

Explanatory Notes

The normal-appearing intermediate cells that are present on a slide provide an appropriate source of comparison for assessing whether nuclear size and appearance meet criteria for ASC-US. Typical ASC-US cells have the size and shape of superficial or intermediate squamous cells. Round or ovoid cells that are approximately one-third the size of superficial cells and therefore resemble large metaplastic or small intermediate cells may also be classified as ASC-US. These cells suggest SIL of intermediate grade (i.e. possible CIN 1 or CIN 2).

The descriptive terms "parakeratosis" and "dyskeratosis" have been used inconsistently, by some to indicate a benign process and by others to indicate an atypical finding. Because these terms are not well defined, they are not part of the formal Bethesda lexicon; the terms are included parenthetically for clarification. Miniature polygonal squamous cells with dense orangeophilic or eosinophilic cytoplasm and small pyknotic nuclei ("parakeratosis") should be classified as NILM if the nuclei appear normal (see Figs. 2.36, 2.37). However, if the nuclei are enlarged, hyperchromatic, or irregular in contour, or if the cells occur in three-dimensional clusters, an interpretation of ASC-US, ASC-H, or SIL should be considered depending on the degree of the abnormality ("atypical parakeratosis")[11] (see Figs. 4.3, 4.17, 5.8, 5.9, 5.23, 5.24). Rarely, the distinction between SIL and decidual cells, trophoblasts, or other unusually encountered elements may also prompt an interpretation of ASC-US.

Determining whether to classify a specimen as NILM or ASC-US may be difficult in the presence of inflammatory or degenerative changes, air-drying with nuclear enlargement, and other artifacts. The patient's age and history should be considered, and previous specimens should be reviewed microscopically if deemed relevant to interpreting the current specimen. Generally, when current cytologic findings favor a reactive process over SIL and the patient has a history of multiple prior negative specimens, the interpretation of NILM should be favored.

Classification of specimens showing mild diffuse nuclear enlargement should take into account the presence of inflammation or infectious agents, adequacy of fixation, patient age, history, and other factors. In general, the presence of cells with pale round nuclei and even chromatin distribution favors an interpretation of NILM over ASC. Most specimens classified as ASC demonstrate a numerically minor subpopulation of atypical cells that are either isolated or occur in small sheets or groupings. ASC may also be an appropriate designation for some specimens that contain abnormal-appearing naked nuclei without associated cytoplasm.

Criteria for ASC-US may differ subtly among laboratories, reflecting differences in stains and techniques for slide preparation.

Atypical Squamous Cells, Cannot Exclude HSIL (ASC-H) (Figs. 4.8–4.16)

ASC-H cells are usually sparse. The following patterns may be seen.

Small Cells with High N/C Ratios: "Atypical (Immature) Metaplasia" (Figs. 4.8–4.13)

Criteria

Cells usually occur singly or in small fragments of less than 10 cells; occasionally, in conventional smears, cells may "stream" in mucus (Fig. 4.9).
Cells are the size of metaplastic cells with nuclei that are about $1^1/_2$ to $2^1/_2$ times larger than normal.
Ratio of nuclear to cytoplasmic (N/C) area may approximate that of HSIL.
In considering a possible interpretation of ASC-H or HSIL, nuclear abnormalities such as hyperchromasia, chromatin irregularity, and abnormal nuclear shapes with focal irregularity favor an interpretation of HSIL.

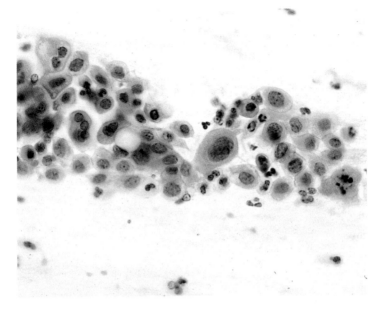

[*] **FIGURE 4.8.** Atypical squamous cells, cannot exclude HSIL (ASC-H) (CP). Premenopausal woman. Loosely cohesive cells with high ratio of nuclear to cytoplasmic area, smooth nuclear contours, and delicate chromatin. Possible interpretations include reactive/reparative metaplastic cells, ASC-H, and high-grade squamous intraepithelial lesion (HSIL). Follow up was CIN 3.

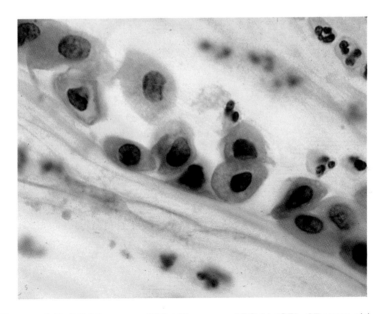

[*] **FIGURE 4.9.** NILM versus ASC-US versus ASC-H (CP). 27-year-old woman, day 8 of menstrual cycle, with history of prior abnormal cytology. Cells with a polygonal shape and dense metaplastic cytoplasm contain enlarged nuclei with even contours, worrisome for an SIL of borderline grade (between LSIL and HSIL).

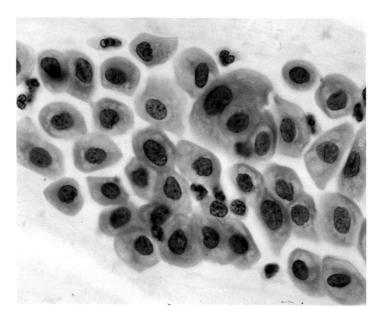

[*] FIGURE 4.10. ASC-H (CP). Cells with metaplastic cytoplasm showing variation in size, shape, and ratio of nuclear to cytoplasmic area.

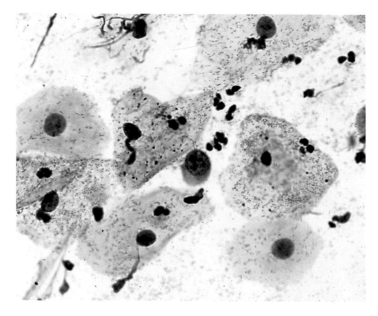

FIGURE 4.11. ASC-H (CP). Single small cell (*center*) with a thin rim of cytoplasm and a nucleus with round contours but irregular chromatin distribution, surrounded by mature squamous cells. Patient was subsequently diagnosed with invasive squamous carcinoma.

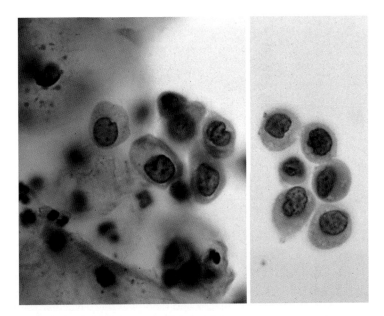

[∗] **Figure 4.12.** ASC-H versus HSIL (LBP). 27-year-old woman. On the *left* are isolated small cells with variable N/C ratios and some cells displaying prominent nuclear irregularity. On the *right* is a high-power view of six small cells with enlarged and irregular, but degenerated, nuclei. Follow up was CIN 3.

"Crowded Sheet Pattern" (Figs. 4.13, 4.14, 4.15)

Criteria

A microbiopsy of crowded cells containing nuclei that may show loss of polarity or are difficult to visualize.

Dense cytoplasm, polygonal cell shape, and fragments with sharp linear edges generally favor squamous over glandular (endocervical) differentiation.

Liquid-Based Preparations

ASC-H cells may appear quite small, with nuclei that are only two to three times the size of the neutrophil nuclei (Fig. 4.14).

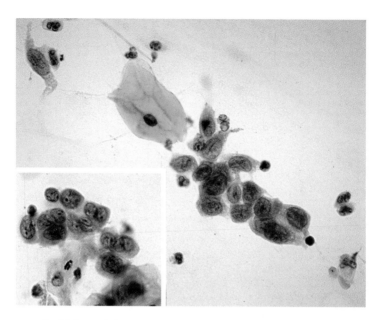

FIGURE 4.13. ASC-H (CP and LBP). Main panel (CP) shows columnar and polygonal cells with dense cytoplasm and high N/C ratio. Nuclear chromatin is granular with small chromocenters or nucleoli. *Insert on lower left corner* contains cells from another specimen (LBP) that appear similar, but possess cytoplasm that is scant and less dense. Possible interpretations include reactive endocervical cells, ASC-H or HSIL. Presence of nucleoli is not typical for HSIL. An HPV test was negative. (© 2001 American Society for Clinical Pathology. Reprinted with permission.)

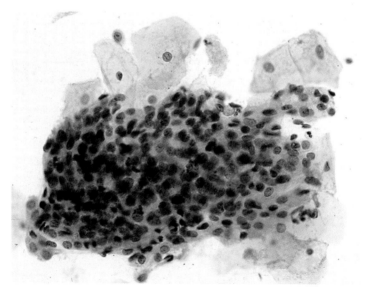

[*] **FIGURE 4.14.** ASC-H (LBP). Thick aggregate of cohesive, overlapping cells containing nuclei with even chromatin and regular borders. The thickness of the cluster makes it difficult to determine if the cells are squamous or glandular. The disorganization of the cells within the group is suggestive of a high grade lesion, however the individual nuclear features are insufficient for a definitive interpretation.

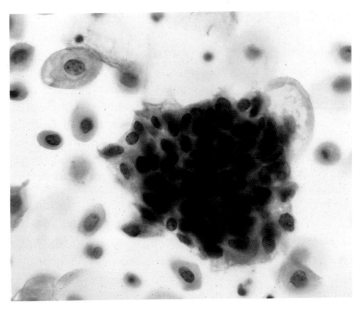

FIGURE 4.15. ASC-H (LBP). 48-year-old woman. A cluster of hyperchromatic cells, the nature of which may be difficult to ascertain in this atrophic specimen. Follow-up was CIN 3.

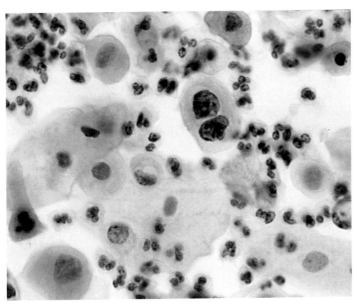

[*] FIGURE 4.16. ASC-H (LBP). Vaginal specimen obtained from patient with prior history of vaginal intraepithelial neoplasia (VAIN) and endometrial carcinoma. Cells with degenerated, markedly hyperchromatic nuclei, worrisome for VAIN, possibly high grade. Follow-up histology was diagnosed as VAIN 3.

Explanatory Notes

Normal metaplastic squamous cells within a specimen may vary considerably in cell size and shape, nuclear size, and nuclear to cytoplasmic ratios. Studies performed on smears have found that immature metaplastic squamous cells have a mean cellular area of about 318 μm^2 as compared to mature metaplastic cells, which have a mean area of 640 μm^2. The mean nuclear area of metaplastic cells is about 50 to 60 μm^2, resulting in a much higher ratio of nuclear to cytoplasmic area than is found in normal intermediate cells. Reserve cells, which are encountered rarely, are even smaller (mean cellular area, 125–175 μm^2), even though they possess nuclei comparable in size to those of metaplastic cells.[10]

When cells with a metaplastic appearance demonstrate relatively mild nuclear enlargement, irregularity, uneven chromatin distribution, or hyperchromasia, HSIL is a concern because the ratio of nuclear to cytoplasmic area may be similar to that found in definite HSIL. The range in size and nuclear appearance of normal metaplastic squamous cells on a slide provide a standard for judging whether cells of concern warrant an interpretation of ASC-H. Defining the range of appearances of normal metaplastic cells in the specimen under review is critical, because this varies with the type of preparation (conventional versus liquid-based) and other factors.

ASC-H may also present as "atypical (immature) metaplasia" in both smears and liquid-based preparations, although this finding is more common in the latter. Note that degenerated nuclei, in the absence of a bona fide SIL, are often irregular or hyperchromatic but the irregularities tend to involve the *entire* nuclear outline, imparting a wrinkled appearance, and the chromatin is smudgy. ASC-H cells are usually sparse; when numerous small atypical cells are identified, the interpretation of HSIL is more likely.

The "crowded sheet pattern" may reflect CIN 2 or CIN 3 (particularly involving endocervical glands), reactive or neoplastic endocervical cells or atrophy with crush artifact[12,13] (see Figs. 5.13, 5.14, 5.31). These cases are sometimes classified as "atypical glandular cells" (AGC), leading to an unexpectedly strong association between the latter category and detection of CIN.[14] Dense cytoplasm, polygonal cell shape, and fragments with sharp linear edges generally favor squamous over glandular differentiation.[15] Excessively vigorous scraping with sampling devices may represent an avoidable cause of thick cell fragments.

Identification of prominent nucleoli is more typical of repair than HSIL; however, nucleoli may be found in cases of HSIL, especially when associated with incipient or established invasion (see Fig. 5.28). Cohesive sheets of cells containing uniform-appearing nuclei with smooth contours and nucleoli favors a reparative process, but nuclear pleomorphism or loss of cohesion may require an interpretation of ASC-H to rule out a lesion.

In atrophic specimens, the small size and high ratio of nuclear to cytoplasmic area typical of atrophic cells may raise concerns about HSIL, especially when nuclear hyperchromasia and smudging associated with degeneration are present (Figs. 4.17, 4.18). Application of topical estrogen may produce sufficient maturation to allow definitive classification of a repeat sample.[16] Blood and inflammation may be present in both atrophic vaginitis and carcinoma; however, frank cellular necrosis would favor a neoplasm. Similar findings may prompt an interpretation of ASC-H following radiation therapy for carcinoma. Typical benign radiated cells show proportionate nuclear and cytoplasmic enlargement associated with cytoplasmic and nuclear degeneration (see Fig. 2.23), but an interpretation of ASC-H is appropriate when a clear distinction from HSIL or carcinoma is impossible. Comparison with the morphology of the original tumor, if available, may help.

Degenerated endometrial cells and macrophages may also possess nuclei that mimic those of HSIL, leading to false-positive interpretations (Figs. 4.19, 4.20). Similarly, some patients wearing an intrauterine device may shed rare small benign cells with an extremely high ratio of nuclear to cytoplasmic area that resemble HSIL. An interpretation of ASC-H or AGC may be appropriate if the etiology of the changes is not certain (see Figs. 2.24, 2.25, 6.5).

In LBPs, ASC-H cells may appear quite small with nuclei that are only two to three times the size of neutrophil nuclei. In some instances, differentiating two overlapping nuclei from a single irregular nucleus may pose difficulties, although this can usually be resolved by focusing up and down at high power (Fig. 4.21). LBPs may also contain cells in the size range of metaplastic cells that possess perfectly round pale nuclei, which nonetheless appear to occupy the majority of the cytoplasm. In some instances, the perception of a high N/C ratio represents an artifact resulting from layering of the cell (squamous metaplastic or endocervical) onto the slide in an orientation that does not demonstrate the total cytoplasmic volume. As mentioned, comparison of nuclear features of the cells in question with normal appearing metaplastic or endocervical cells is useful.

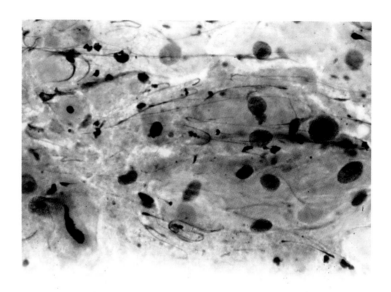

[*] Figure 4.17. ASC-H (CP). 50-year-old postmenopausal woman with prior abnormal cytology. Two cells with extremely hyperchromatic, degenerated nuclei, and orangeophilic cytoplasm, in a background of atrophy with lysed cells and debris. Follow-up after treatment with estrogen to induce maturation demonstrated CIN 2.

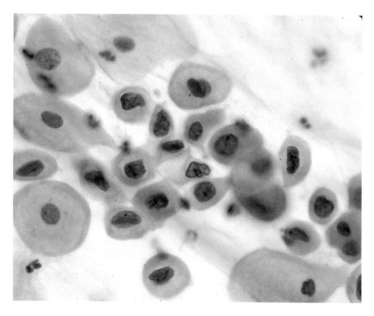

[*] Figure 4.18. ASC-H (CP). Smear from postmenopausal patient containing ovoid cells with irregular poorly preserved nuclei. Possible interpretations include NILM (atrophy), ASC-H and HSIL. Topical estrogen with follow-up may clarify the significance of these cells.

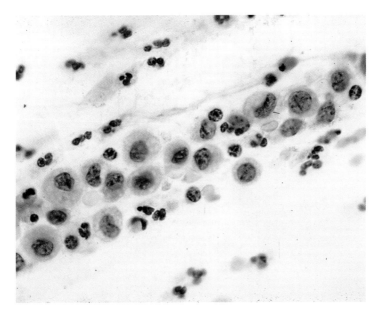

[*] **Figure 4.19.** NILM, histiocytes (CP). Streaming pattern of single cells with round, ovoid, and bean-shaped nuclei. Cells possess fine cytoplasmic vacuoles that may resemble degenerative vacuoles sometimes found in normal metaplasia, ASC-H, and HSIL. By contrast, cells of squamous lineage typically are polygonal in shape and possess dense cytoplasm.

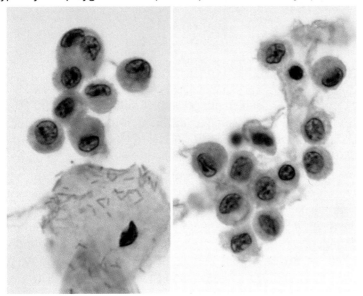

Figure 4.20. NILM, histiocytes (LBP). Routine screen from 32-year-old woman. Cells possess eccentric oval and round nuclei and foamy cytoplasm. The rounder shape of most cells in LBP as compared to CP may lead to uncertainty about the cell type; however, definitive assessment is usually possible under high power examination. Both sides of this image show the cytologic appearance of histiocytes. Follow up was NILM.

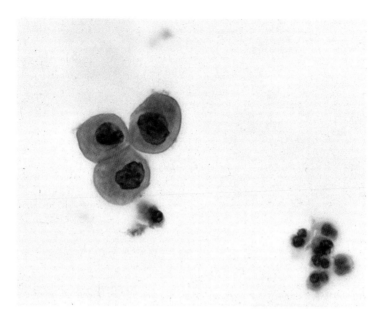

FIGURE 4.21. NILM versus ASC-H (LBP). Ovoid cells with dense cytoplasm and findings that suggest either nuclear irregularity or binucleation. Focusing up and down may clarify the interpretation.

Management

The new consensus management guidelines tailored to the 2001 Bethesda System classification recommend different follow-up for women with ASC-US as opposed to those with ASC-H.[16] Oncogenic (high-risk) HPV DNA testing is the preferred management for ASC-US when it can be performed concurrently with cytology; repeat cytologic testing and immediate colposcopy also represent acceptable management. In contrast, the recommended management for ASC-H is colposcopy. Management of women with ASC-H following colposcopy that does not result in a histologic diagnosis of CIN 2 or a more severe lesion should be individualized based on review of all pathologic and clinical findings. In contradistinction to definitive interpretations of HSIL, immediate treatment without colposcopy is not acceptable. Careful review is required before treating a woman with ASC-H who does not have histologically diagnosed CIN 2 or worse.

Results from the ASCUS/LSIL Triage Study (ALTS) found that the interpretations of ASC-H were associated with a higher risk of oncogenic HPV DNA detection and greater risk of underlying CIN 2 or worse (30%–40%) compared to ASC-US (10%–15% risk of CIN 2 or worse). These risk associations were similar for both conventional smears and liquid-based preparations.[5,8] Laboratories that routinely use oncogenic HPV testing for managing women with ASC-US are encouraged to compare virologic testing results, cytomorphology, and available follow-up. The percentage of HPV infections among women with ASC-US is strongly correlated with age,[17] but results among cytologists within a laboratory should be similar, given comparable caseloads. When histopathologic CIN 3 is present in loop electrosurgical excision procedure (LEEP) specimens obtained following a cytologic report of ASC, the extent of CIN 3 is generally limited.[18] Identification of CIN 3 in multiple blocks of a LEEP specimen following a cytologic report of ASC should prompt a quality assurance review including rescreening of the cytologic slide when possible. Laboratories are also encouraged to monitor follow-up of ASC-US and ASC-H separately when feasible to extend our understanding of these categories based on research findings to clinical practice.

Sample Reports

Example 1:
   Epithelial Cell Abnormality:
   Atypical squamous cells–undetermined significance (ASC-US)
   Comment:
   Suggest high-risk HPV testing if clinically warranted (if reflex testing
      not ordered or if conventional preparation)
   OR
   Specimen sent for reflex HPV testing per clinician request.

Example 2:
   Epithelial Cell Abnormality:
   Atypical squamous cells–cannot exclude a high-grade squamous intraepithelial lesion (ASC-H).
   Comment: Suggest colposcopy/biopsy as clinically indicated.

   For examples of reporting ASC-US in conjunction with HPV testing,
      see Chapter 9 on Ancillary Testing.

Bethesda System 2001 Workshop Forum Group Moderators:

Mark E. Sherman, M.D., Fadi W. Abdul-Karim, M.D., Jonathan Berek, M.D., Patricia Braly, M.D., Robert Gay, C.T. (ASCP), Celeste Powers, M.D., Ph.D., Mary K. Sidawy, M.D., and Sana O. Tabbara, M.D.

References

1. Solomon D, Davey D, Kurman R, et al. The 2001 Bethesda System: terminology for reporting results of cervical cytology. *JAMA* 2002;287:2114–2119.
2. National Cancer Institute Workshop. The 1991 Bethesda System for reporting cervical/vaginal cytologic diagnoses: report of 1991 Bethesda Workshop. *JAMA* 1992; 267:892.
3. Davey DD, Woodhouse S, Styer P, et al. Atypical epithelial cells and specimen adequacy. Current laboratory practices of participants in the College of American Pathologists Interlaboratory Comparison Program in Cervicovaginal Cytology. *Arch Pathol Lab Med* 2000;124:203–211.
4. Kurman J, Henson DE, Herbst AL, et al. The 1992 National Cancer Institute Workshop. Interim guidelines for management of abnormal cervical cytology. *JAMA* 1994;271:1866–1869.
5. The ALTS Group: Results of a randomized trial on the management of cytology interpretations of atypical squamous cells of undetermined significance. *Am J Obstet Gynecol* 2003;183:1383–1392.
6. Pitman MB, Cibas ES, Powers CN, et al. Reducing or eliminating use of the category of atypical squamous cells of undetermined significance decreases the diagnostic accuracy of the Papanicolaou smear. *Cancer Cytopathol* 2002;96:128–134.
7. Quddus MR, Sung CJ, Steinhoff MM, et al. Atypical squamous metaplastic cells: reproducibility, outcome, and diagnostic features on ThinPrep Pap test. *Cancer Cytopathol* 2001;93:16–22.
8. Sherman ME, Solomon D, Schiffman M (for the ALTS Group). Qualification of ASCUS. A comparison of equivocal LSIL and equivocal HSIL cervical cytology in the ASCUS LSIL Triage Study. *Am J Clin Pathol* 2001;116:386–394.
9. Stoler MH, Schiffman M (for the ALTS Group). Interobserver reproducibility of cervical cytologic and histologic interpretations: realistic estimates from the ASCUS LSIL Triage Study. *JAMA* 2001;285:1500–1505.
10. Patten SF Jr. Benign proliferative reactions and squamous atypia of the uterine cervix. In: Wied GL, Bibbo M, Keebler CM, Koss LG, Pattern SF, Rosenthal DL, eds. *Compendium on Diagnostic Cytology*, 8th Ed. Chicago: Tutorials of Cytology, 1997:81–85.
11. Abramovich CM, Wasman JK, Siekkinen P, et al. Histopathologic correlation of atypical parakeratosis diagnosed on cervicovaginal cytology. *Acta Cytol* 2003;47(3): 405–409.
12. Boon ME, Zeppa P, Ouwerkerk-Noordam E, et al. Exploiting the "toothpick effect" of the Cytobrush by plastic embedding of cervical samples. *Acta Cytol* 1991;35:57–63.
13. Drijkoningen M, Meertens B, Lauweryns J. High grade squamous intraepithelial le-

sion (CIN 3) with extension into the endocervical clefts. Difficulty of cytologic differentiation from adenocarcinoma in situ. *Acta Cytol* 1996;40:889–894.

14. Veljovich DS, Stoler MH, Andersen WA, et al. Atypical glandular cells of undetermined significance: a five-year retrospective histopathologic study. *Am J Obstet Gynecol* 1998;179:382–390.

15. Ronnett BM, Manos MM, Ransley J, et al. Atypical glandular cells of undetermined significance (AGUS): cytopathologic features, histopathologic results and human papillomavirus DNA detection. *Hum Pathol* 1999;30:816–825.

16. Wright TC Jr, Cox JT, Massad LS, et al. 2001 Consensus guidelines for the management of women with cervical cytological abnormalities. *JAMA* 2002;287:2120–2129.

17. Sherman ME, Schiffman M, Cox JT, for the ALTS Study Group. Effects of age and human papilloma viral load on colposcopy triage: data from the randomized Atypical Squamous Cells of Undetermined Significance/Low-Grade Squamous Intraepithelial Lesion Triage Study (ALTS). *J Natl Cancer Inst* 2002;94:102–107.

18. Sherman ME, Wang SS, Tarone R, et al. Histopathologic extent of cervical intraepithelial neoplasia 3 lesions in the Atypical Squamous Cells of Undetermined Significance/Low-grade Squamous Intraepithelial Lesion Triage Study: implications for subject safety and lead-time bias. *Cancer Epidemiol Biomarkers Prev.* 2003;12:372–379.

Chapter 5

Epithelial Cell Abnormalities: Squamous

Thomas C. Wright, Rose Marie Gatscha, Ronald D. Luff, and Marianne U. Prey

Epithelial Cell Abnormalities:

Squamous Cell

- ❏ Squamous Intraepithelial Lesion (SIL)
  - ➤ Low-grade squamous intraepithelial lesion (LSIL)
  - ➤ High-grade squamous intraepithelial lesion (HSIL)
    - • with features suspicious for invasion (*if invasion is suspected*)
- ❏ Squamous cell carcinoma

Background

Squamous intraepithelial lesion (SIL) encompasses the spectrum of noninvasive cervical squamous epithelial abnormalities associated with human papillomavirus (HPV), which range from the cellular changes that are associated with transient HPV infection to abnormal cellular changes representing high-grade precursors to invasive squamous cancer. In The Bethesda System (TBS), this spectrum is divided into low-grade (LSIL) and high-grade (HSIL) categories. Low-grade lesions encompass the cellular changes variously termed "HPV cytopathic effect" (koilocytosis) and mild dysplasia or cervical intraepithelial neoplasia (CIN) 1. High-grade lesions encompass moderate dysplasia, severe dysplasia, and carcinoma in situ or CIN 2, 3.

HPV-associated squamous cell changes or cytopathic effects were first described as perinuclear clearing with peripheral condensation of the cytoplasm.[1,2] These features, termed "koilocytosis," were initially considered to represent a process separate from "true" dysplasia. However, mounting evidence over the past 20 years has established HPV as the main causal factor in the pathogenesis of virtually all cervical cancer precursors and cancers.[3] HPV-DNA is detected by molecular techniques in the vast majority of SILs and cancers. The majority (>98%) of invasive cervical cancers and their precursors contain HPV types referred to as "high-risk" HPVs, the most common being HPV 16.[4] High-risk HPVs are associated with both low- and high-grade SIL, but are observed with significantly greater frequency in the high-grade group.[5]

Conceptually, HPV-associated abnormalities can be divided into transient infections that generally regress over the course of 1 to 2 years[6] and HPV persistence that is associated with an increased risk of developing a cancer precursor or invasive cancer.[7-9] There has been a shift in the United States in recent years with regard to the management of biopsy-confirmed CIN 1 based on the recognition that most CIN 1, especially in young women, represents a self-limited HPV infection. The current emphasis of cervical cancer screening is focused on detection and treatment of biopsy-confirmed high-grade disease, regardless of the distinction between CIN 2 and CIN 3.[10]

At the 1988 Bethesda workshop, the spectrum of SIL was subdivided into two categories based on (1) the desire to use morphologic categories that relate to the biology and clinical management of HPV-associated lesions, and (2) the acknowledged low inter- and intraobserver reproducibility with conventional three- and four-grade classification systems.[11,12] It has been argued that a two-tiered system provides less information to clinicians than a three-tiered CIN terminology.[13] However, the cytologic distinction of CIN 2 and CIN 3 is poorly reproducible, and combining the cytologic correlates of biopsy-confirmed CIN 2 and CIN 3 into a single HSIL category was shown, in the ASCUS LSIL Triage Study (ALTS), to have improved reproducibility (M. Schiffman, personal communication).

Another concern voiced about the two-tiered classification is that the dividing line between low-grade and high-grade precursors should be set between CIN 2 and CIN 3 because the natural history of untreated CIN 2 is closer to that of CIN 1 than it is to CIN 3.[14] In some European countries CIN 1 and CIN 2 are grouped together for treatment purposes.[13] However, as a screening test, cervical cytology must emphasize sensitivity. Given the variability in the interpretation and biologic behavior of "cytologic CIN2",[15] setting the cytologic threshold for low-grade and high-grade lesions between CIN 1 and CIN 2 is considered appropriate. This cut point also demonstrated the best interobserver reproducibility using a dichotomous positive/negative result, based on data from ALTS (M. Schiffman, personal communication).

Even with only two categories of SIL, there is an overall 10% to 15% interpathologist discrepancy rate between LSIL and HSIL interpretations on cervical cytology slides.[16] Cytology may also be discrepant with histology; 15% to 25% of women with LSIL cytology are found to have histologic CIN 2 or CIN 3 upon further evaluation.[17]

Low-Grade Squamous Intraepithelial Lesion (LSIL) (Figs. 5.1–5.11)

Squamous cell changes associated with HPV infection encompass "mild dysplasia" and "CIN 1." Several studies have demonstrated that the morphologic criteria for distinguishing "koilocytosis" from mild dysplasia or CIN I vary among investigators and lack clinical significance. In addition, both lesions share similar HPV types, and their biologic behavior and clinical management are similar, thus supporting a common designation of LSIL.[18-20]

Criteria

Cells occur singly and in sheets.

Cytologic changes are usually confined to cells with "mature" or superficial-type cytoplasm.

Overall cell size is large, with fairly abundant "mature" well-defined cytoplasm.

Nuclear enlargement more than three times the area of normal intermediate nuclei results in a slightly increased nuclear to cytoplasmic ratio.

Variable degrees of nuclear hyperchromasia are accompanied by variations in nuclear size, number, and shape.

Binucleation and multinucleation are common (Fig. 5.2).

Chromatin is often uniformly distributed, but coarsely granular; alternatively, the chromatin may appear smudged or densely opaque (Figs. 5.1, 5.2).

Nucleoli are generally absent or inconspicuous if present.

Contour of nuclear membranes is often slightly irregular, but may be smooth (Fig. 5.1).

Cells have distinct cytoplasmic borders.

Perinuclear cavitation ("koilocytosis"), consisting of a sharply delineated clear perinuclear zone and a peripheral rim of densely stained cytoplasm, is a characteristic feature but is not required for the interpretation of LSIL (Fig 5.1); alternatively, the cytoplasm may appear dense and orangeophilic (keratinized).

Cells with cytoplasmic perinuclear cavitation or dense orangeophilia must also show nuclear abnormalities to be diagnostic of LSIL (Figs. 5.3, 5.4, 5.5); perinuclear halos in the *absence of nuclear abnormalities* do not qualify for the interpretation of LSIL (Fig. 5.7).

Liquid-Based Preparations

Enlarged nuclei may not show significant nuclear hyperchromasia (Fig. 5.5).

Angulated clusters of atypical/dysplastic cells may be more clearly visualized; they should be classified based on the degree of nuclear abnormality (Fig. 5.9).

Explanatory Notes

An interpretation of LSIL should be based on strict criteria to avoid over-interpretation and unnecessary treatment of women for nonspecific morphologic changes. For example, cytoplasmic perinuclear halos without accompanying atypical nuclear features should not be considered LSIL (Fig. 5.7). Terms such as "koilocytosis," "koilocytotic atypia," and "condylomatous atypia" are not included in TBS terminology. So-called "atypical parakeratosis" showing nuclear abnormalities should be categorized as SIL (Fig. 5.8). Specimens with borderline changes that fall short of a definitive SIL interpretation may be categorized as "Atypical squamous cells–of undetermined significance" (ASC-US) (Figs. 5.6, 5.10, 5.11).

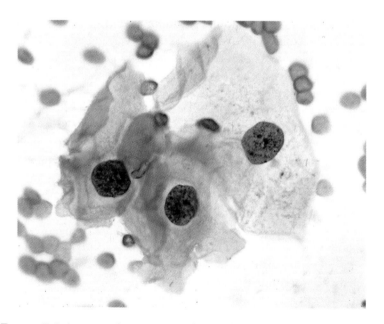

[∗] FIGURE 5.1. Low-grade squamous intraepithelial lesion (LSIL) (conventional preparation, CP). Nuclear enlargement and hyperchromasia is of sufficient degree for the interpretation of LSIL. HPV-associated cytoplasmic changes are not a prerequisite for LSIL. [[∗] Bethesda Interobserver Reproducibility Project (BIRP) image (see xviii Introduction).]

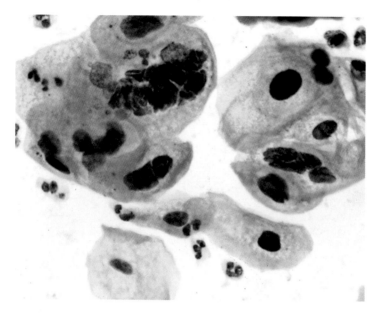

[*] **FIGURE 5.2.** LSIL (liquid-based preparation, LBP). A 32-year-old woman, day 15, routine cervical cytology screening. Note the overall large cell size, "smudged" nuclear chromatin, well-defined cytoplasm, and multinucleation.

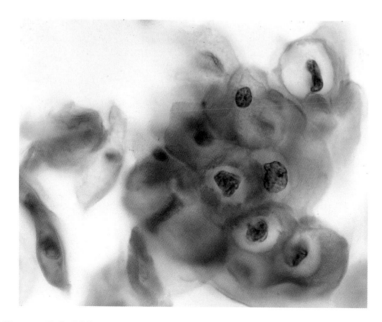

[*] **FIGURE 5.3.** LSIL (LBP). 26-year-old woman with prior abnormal cytology. Nuclear features are consistent with LSIL. HPV cytopathic effect or "koilocytosis" is also seen.

[*] **FIGURE 5.4.** LSIL (LBP). Routine screen from 32-year-old woman. Note that, in addition to perinuclear cavitation, nuclear abnormalities as seen here are required to make an interpretation of LSIL.

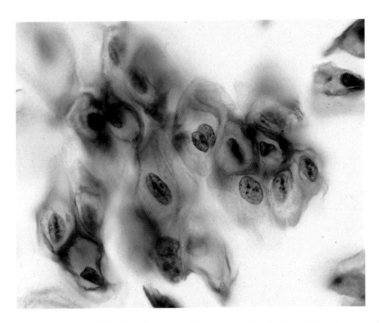

[*] **FIGURE 5.5.** LSIL (LBP). 22-year-old woman. Nuclear hyperchromasia may not be as obvious in LBPs as compared to conventional smears; however, other criteria for interpretation of SIL are present. Follow up was CIN 1.

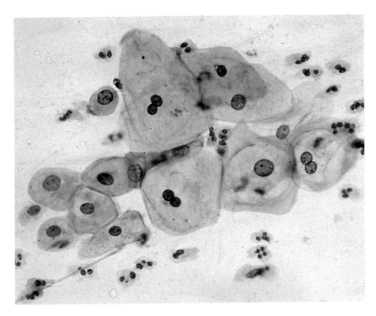

FIGURE 5.6. ASC-US versus LSIL (CP). Binucleation and subtle koilocytosis are seen but without significant nuclear changes. Cell changes are border-line between ASC-US and LSIL.

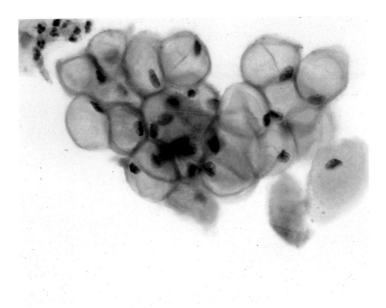

[*] **FIGURE 5.7.** NILM (CP). Routine screen 24-year-old woman. Negative for intraepithelial lesion or malignancy (NILM) (LBP). Glycogen in squamous cells can give the appearance of "pseudokoilocytosis." Nuclear abnormalities required for an interpretation of LSIL are absent. Follow-up was NILM.

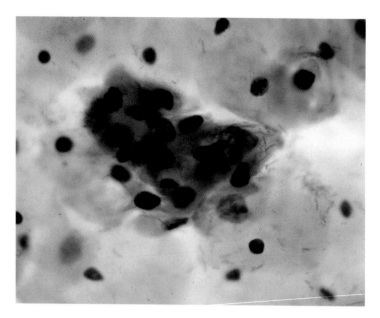

[*] **Figure 5.8.** SIL, grade cannot be determined (CP). Atypical squamous cells with orangeophilic cytoplasm ("atypical parakeratosis"); interpretation of SIL is based on nuclear features. The cells are indicative of an SIL, however such lesions may be difficult to grade.

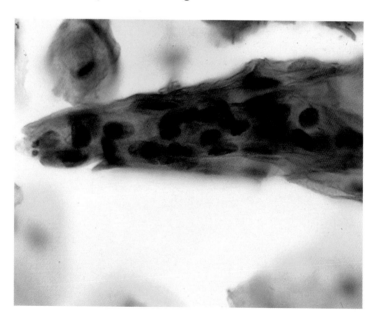

[*] **Figure 5.9.** ASC-US versus LSIL (LBP). 32-year-old woman. Clusters of squamous cells may be seen in "spike-like" aggregates; such clusters should be classified based on the degree of nuclear abnormalities. This patient had an LSIL interpretation on a conventional smear 2 months before this LBP which was interpreted as ASC-US ("atypical parakeratosis").

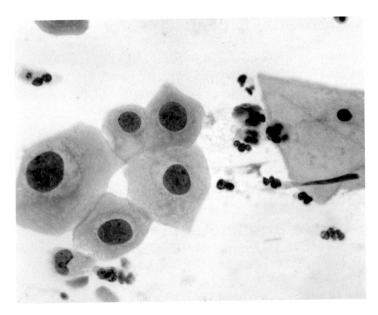

[*] FIGURE 5.10. ASC-US versus LSIL (CP). Nuclear features are borderline between those required for ASC-US and LSIL.

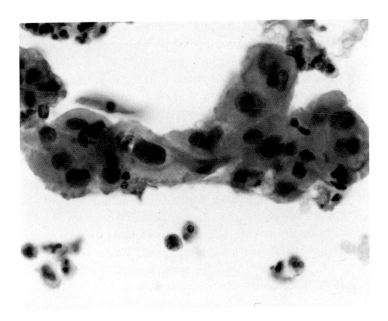

FIGURE 5.11. ASC-US versus LSIL (LBP). 18-year-old woman. Nuclear features are borderline between those required for ASC-US and LSIL. Follow up was CIN 1.

Management of LSIL

In data from the ASCUS/LSIL Triage Study (ALTS), high-risk HPV types were detected in 85% of LSIL cases, with the conclusion being that HPV testing is not a useful triage strategy.[21] The 2001 ASCCP consensus guidelines recommend colposcopy for the initial management of LSIL.[10] HPV testing may have a potential role in the detection of persistent high-risk-type viral infection in patients with SIL, especially because evidence has accumulated that persistent HPV infection is the major risk factor for "progression."[22]

High-Grade Squamous Intraepithelial Lesion (HSIL) (Figs. 5.12–5.35)

Criteria

Cytologic changes affect cells that are smaller and less "mature" than the cells in LSIL.

Cells occur singly, in sheets, or in syncytial-like aggregates (Figs. 5.12, 5.13).

Hyperchromatic clusters should be carefully assessed (Fig. 5.14).

Overall cell size is variable, and ranges from cells that are similar in size to those observed in LSIL to quite small basal-type cells.

Nuclear hyperchromasia is accompanied by variations in nuclear size and shape (Fig. 5.15).

Degree of nuclear enlargement is more variable than that seen in LSIL. Some HSIL cells have the same degree of nuclear enlargement as in LSIL, but the cytoplasmic area is decreased, leading to a marked increase in the nuclear/cytoplasmic ratio (Figs. 5.17, 5.18). Other cells have very high nuclear/cytoplasmic ratios, but the actual size of the nuclei may be considerably smaller than that of LSIL (Fig. 5.19).

Chromatin may be fine or coarsely granular and evenly distributed.

Contour of the nuclear membrane is quite irregular and frequently demonstrates prominent indentations (Figs. 5.17, 5.20) or grooves (Fig. 5.22).

Nucleoli are generally absent, but may occasionally be seen, particularly when HSIL extends into endocervical gland spaces (Fig. 5.21).

Appearance of cytoplasm is variable; it can appear "immature," lacy, and delicate (Fig. 5.22) or densely metaplastic (Fig. 5.18); occasionally the cytoplasm is "mature" and densely keratinized (keratinizing HSIL) (Fig. 5.24).

Liquid-Based Preparations

Dispersed abnormal single cells are seen more often than sheets and syncytial aggregates. Isolated cells may be present in the empty spaces between cell clusters (Fig. 5.25).

Relatively fewer abnormal cells may be present.

As in LSIL, in liquid-based preparations (LBPs) some HSIL cells may not show significant hyperchromasia, but other cytologic features of HSIL (high nuclear to cytoplasmic ratio and irregular nuclear membrane) are present (Figs. 5.20, 5.25).

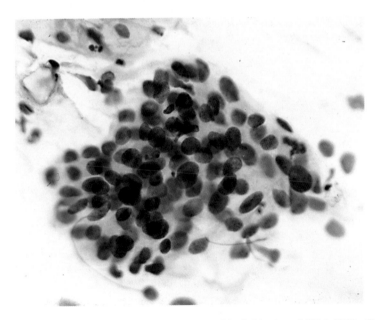

FIGURE 5.12. High-grade squamous intraepithelial lesion (HSIL) (CP). The dysplastic cells are seen here in a syncytial cluster or hyperchromatic crowded group.

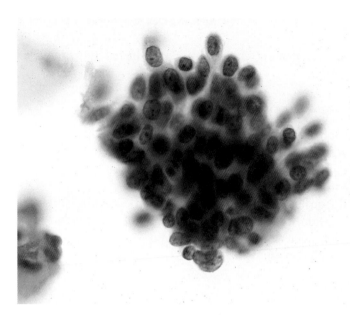

FIGURE 5.13. HSIL—syncytial cluster (LBP). As in conventional smears, crowded hyperchromatic cell groups should be examined with care. If a squamous abnormality is suspected, a thorough search for single dysplastic cells in the background is warranted. Follow up was CIN 3 with endocervical gland involvement.

FIGURE 5.14. HSIL (CP). 58-year-old postmenopausal woman on hormone replacement therapy. Hyperchromatic crowded groups seen at low power require careful examination at higher magnification. Flattening at the edge of the cell cluster and whorling in the center are suggestive of HSIL over a glandular abnormality. Follow up was CIN 3 with endocervical gland involvement.

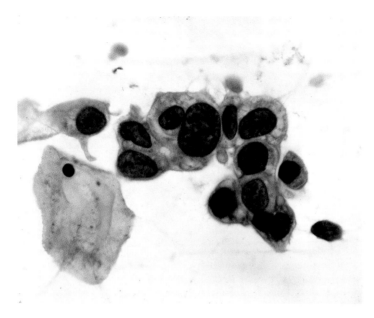

FIGURE 5.15. HSIL (CP). Note variation in nuclear size and shape and delicate cytoplasm.

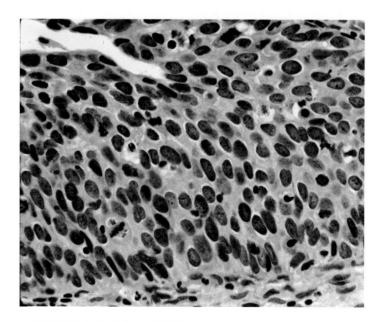

FIGURE 5.16. CIN 3 (histology, H&E).

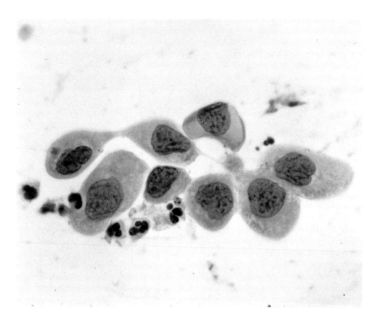

FIGURE 5.17. HSIL (CP). Nuclear changes are HSIL; however, the nuclear/cytoplasmic (N/C) ratio is on the low end for HSIL. For those using CIN, this would be considered CIN 2.

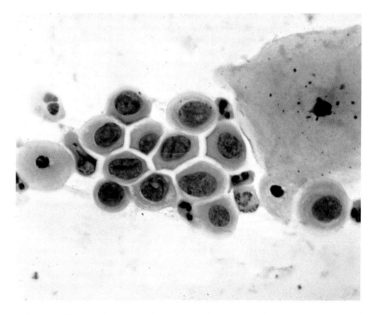

FIGURE 5.18. HSIL (CP). Note the "metaplastic" or dense cytoplasm, in contrast to that seen in Figure 5.15.

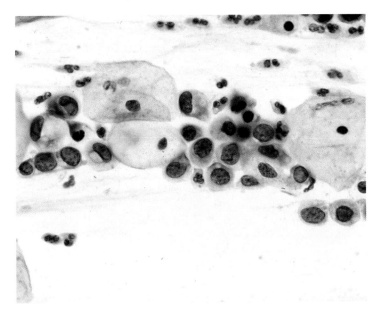

[*] **FIGURE 5.19.** HSIL (CP). HSIL cells with some variation in cell size and N/C ratios. A cluster such as this may be misinterpreted as squamous metaplastic cells if examined only under lower magnification. Follow up was CIN 3.

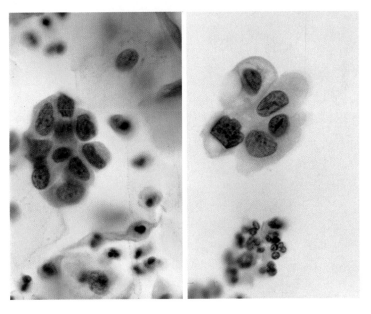

[*] **FIGURE 5.20.** HSIL (LBP). Note the nuclear envelope irregularities and abnormally distributed chromatin. In LBPs, hyperchromasia may not be as prominent as in conventional smears.

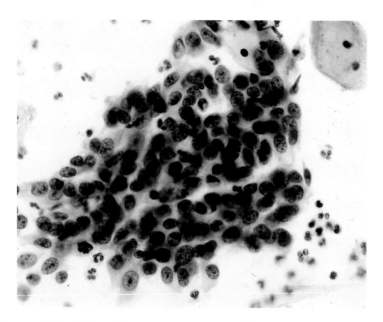

[*] FIGURE 5.21. HSIL (CP). 42-year-old woman. Although uncommon, nucleoli may be seen in HSIL, especially with extension into endocervical gland spaces. The chromatin may appear less coarsely granular.

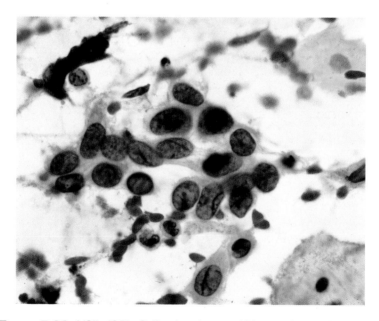

[*] FIGURE 5.22. HSIL (CP). Cells showing ovoid hyperchromatic nuclei with grooves and scant cytoplasm with ill defined borders.

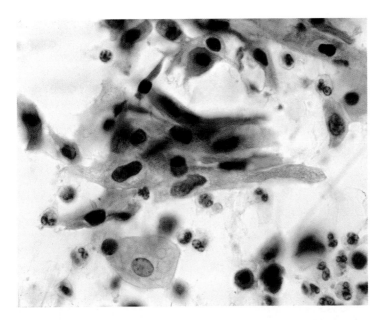

[*] FIGURE **5.23.** SIL, grade cannot be determined (CP). Classification of atypical keratinized cells depends on the degree of nuclear abnormality, the N/C ratio, and to some extent on pleomorphism of cell shape (see also Figs. 5.8, 5.24, and 5.39).

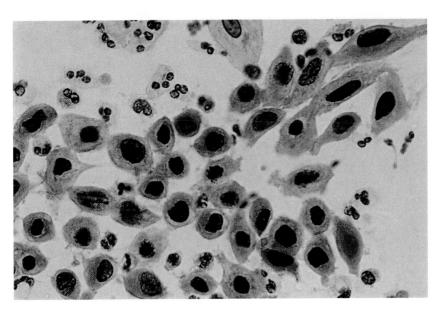

FIGURE **5.24.** HSIL—"keratinizing lesion" (CP). The criteria of N/C ratio and degree of nuclear abnormalities used for grading SIL may be more difficult to apply to keratinizing lesions. The extent of abnormality here qualifies for an interpretation of HSIL (contrast with Fig. 5.8 and 5.23).

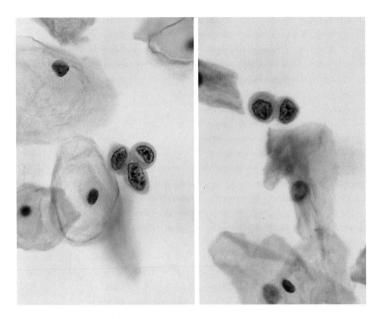

[*] Figure 5.25. HSIL (LBP). 29-year-old woman in high risk clinic. Close attention to isolated cells is required when screening LBPs because the abnormal isolated cells may not be as apparent as clusters of HSIL cells and may lie between benign cell clusters or in "empty spaces" on the preparation. When the criteria for HSIL are met, such cells should be interpreted as HSIL and not ASC-H. Both images (*right and left*) demonstrate such cells. Follow up was CIN 3.

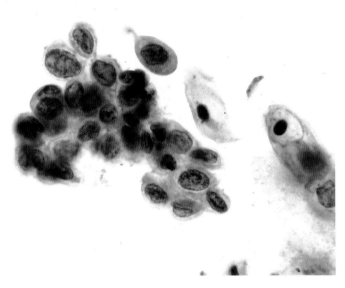

[*] Figure 5.26. HSIL (LBP). Cluster of HSIL cells with high N/C ratio and nuclear abnormalities.

Explanatory Notes

SILs of Indeterminate Grade (Fig. 5.27)

Cells may have cytologic features that lie between low- or high-grade SIL (Fig. 5.27). Although occasional "borderline" cases occur, attention to morphologic features usually supports classification as either LSIL or HSIL. LSIL is characterized by nuclear enlargement at least three times the area of a normal intermediate cell nucleus. Although the nucleus is hyperchromatic, the chromatin is distributed uniformly or it may appear degenerated. Features that favor a high-grade lesion include higher nuclear/cytoplasmic ratios, greater irregularities in the outline of the nuclear envelope, coarsening of nuclear chromatin, and chromatin clumping. The appearance of the cytoplasm also may assist in determining whether a "borderline" case is low- or high-grade SIL; LSIL typically involves squamous cells with "mature," intermediate, or superficial-type cytoplasm with well-defined polygonal cell borders. Cells of an HSIL have a more "immature" type of cytoplasm, which be can either lacy and delicate or dense/metaplastic, with rounded cell borders. Overall, cell size is smaller

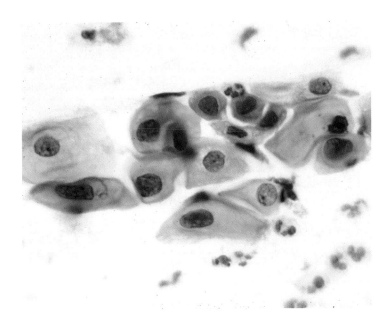

[*] FIGURE **5.27.** SIL, grade cannot be determined (CP). Routine screen from 28-year-old woman. Most of these cells qualify as LSIL; there are three atypical metaplastic cells at top center that raise the possibility of a high grade lesion. Follow up was CIN 2.

in HSIL as compared with LSIL. In occasional cases where it is not possible to grade an SIL as clearly low or high,[23] an interpretation of "SIL, grade cannot be determined" may be appropriate (See Figs. 5.23, 5.27). In other cases, definite "LSIL but with rare cells suggestive of HSIL" may be identified and reported as such. A designation of "atypical squamous cells, cannot exclude HSIL" (ASC-H) is appropriate for specimens with some features of HSIL but that fall short of a definitive interpretation of SIL (see Figs. 4.10–4.18).

Keratinizing Lesions (Figs. 5.23, 5.24)

Although most HSILs are characterized by cells with a high nuclear/cytoplasmic ratio, some high-grade lesions are composed of cells with more abundant, but abnormally keratinized, cytoplasm (see Figs. 5.23, 5.24, 5.28). These cells may be shed singly or in three-dimensional clusters and have enlarged hyperchromatic nuclei, often with dense chromatin that obscures other nuclear features. In addition, these cells are often quite pleomorphic with marked variation of nuclear size (anisokaryosis) and cellular shape, including elongate, spindle, caudate, and tadpole cells. In contrast to invasive squamous carcinoma, nucleoli and tumor diathesis are generally absent. Such lesions have been variously termed "atypical condyloma," "keratinizing dysplasia," and "pleomorphic dysplasia." At times, these keratinized lesions may be indistinguishable from invasive carcinoma, especially in samples with a relatively scant number of abnormal cells. In these instances, an explanatory note may be appended to indicate that the differential diagnosis includes an invasive squamous cell carcinoma, or the interpretation of *HSIL with features suspicious for invasion* can be used (Fig. 5.28).

SIL with Gland Involvement (Figs. 5.29–5.33)

When SIL, especially HSIL, extends into the endocervical glands, the cell clusters may be misinterpreted as being glandular in origin. Clues that the lesion is squamous in nature include centrally located cells showing spindling or whorling (Fig. 5.14) with flattening of the nuclei at the periphery of the cluster, giving a smooth, rounded border (Figs. 5.29, 5.30). However, HSIL in glands may demonstrate peripheral palisading of cells and nuclear pseudostratification, features that are usually associated with adenocarcinoma in situ (AIS).[24] Attention to nuclear features may help because the chromatin pattern in squamous dysplasia is not as coarsely granular as in AIS (Fig. 6.21).

On LBPs, loss of central cell polarity and piling within cell groups is observed in HSIL involving glands but not in AIS. Also, in contrast to conventional smears, nucleoli may be visualized in HSIL within glands on LBPs, but are not as prominent as in AIS.[25] HSIL and AIS can also coexist (Fig. 6.33).

[*] **FIGURE 5.28.** HSIL (LBP). 42-year-old woman. Keratinized dysplastic cells with nucleoli, and angulated or "carrot"-shaped nuclei that may raise suspicion for invasion and qualify for an interpretation of HSIL, cannot rule out invasion. Follow up showed only CIN 3 (keratinizing).

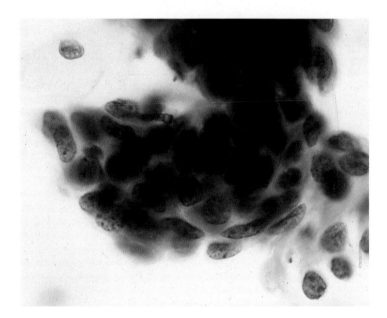

FIGURE 5.29. HSIL with extension into gland space (LBP). Note flattening of cells at the edge of the cluster, a feature that favors HSIL over a glandular lesion.

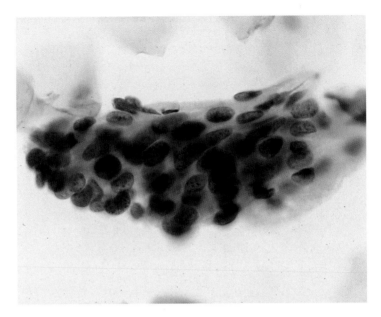

FIGURE 5.30. HSIL (LBP). 44-year-old woman. Syncytial cluster of HSIL cells with features of endocervical gland extension. Such "hyperchromatic crowded groups" may raise a wide differential diagnosis under low magnification; attention to architectural pattern and cellular detail are necessary for correct interpretation. Follow-up showed CIN 3 with endocervical gland involvement.

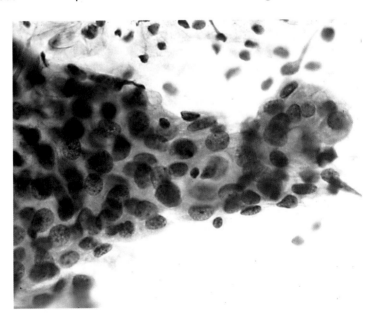

FIGURE 5.31. HSIL (CP). 30-year-old woman with prior atypical glandular cells on cytology. When HSIL lesions involve endocervical glands, they may show features that overlap with those of adenocarcinoma in situ (AIS). Note normal columnar cells with residual mucin at the *right upper edge* of the cell cluster. Follow-up showed CIN with endocervical gland involvement.

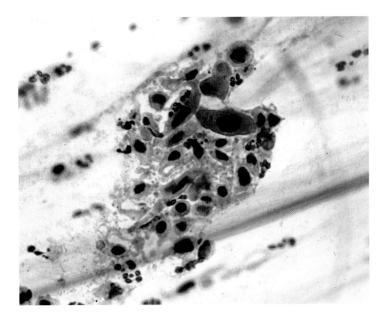

FIGURE 5.32. HSIL with features suspicious for invasion (CP). 71-year-old postmenopausal woman. HSIL filling endocervical glands may undergo focal necrosis that can mimic the tumor diathesis associated with invasive lesions. Follow up showed CIN 3 extending into glands with focal epithelial necrosis, but no invasion.

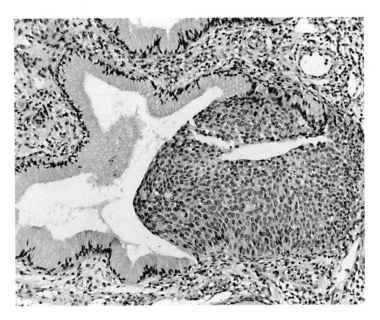

FIGURE 5.33. CIN 3 with extension into endocervical glands (histology, H&E).

Patterns of HSIL That Can Be Missed

Rare Small HSIL Cells (Fig. 5.25)

Specimens with rare, small, high nuclear / cytoplasmic ratio HSIL cells may be problematic with regard to identifying the cells as well as categorizing the abnormality.[26] There is a higher probability of a false-negative result on conventional cytology when there are relatively few detached neoplastic cells or when only a few large groups of neoplastic cells are present.[27] LBPs frequently have fewer diagnostic cells compared to conventional smears, although the cells may be better visualized. If rare abnormal cells are identified but the findings fall short of an interpretation of HSIL, the specimen may be reported as ASC-H (see Figs. 4.16, 4.17). The differential diagnosis of isolated cells with high nuclear/cytoplasmic ratio includes immature squamous metaplasia, cellular changes associated with intrauterine device (IUD) use (see Figs. 2.25, 6.5), and isolated cells of endocervical or endometrial origin.

Stream of HSIL Cells, Usually Within Mucus (Figs. 5.34, 5.35)

HSIL in mucus strands can resemble histiocytes/superficial endometrial stromal cells or degenerated endocervical cells as in microglandular hyperplasia (Figs. 5.34, 5.35). The low-power pattern of small cells in a streak of mucus warrants evaluation at higher power. This pattern is generally not observed in liquid-based preparations which fragment mucus.

HSIL with Features Suspicious for Invasion (Figs. 5.28, 5.32)

In rare cases of HSIL, invasive carcinoma is difficult to exclude. This situation may occur when there are highly pleomorphic HSIL cells with keratinized cytoplasm present that are not accompanied by the characteristic background features of invasion (necrosis or tumor diathesis; see Fig. 5.28). Conversely, there may be features suggesting tumor diathesis (blood, necrosis, or granular proteinaceous debris in the background) but malignant cells are not identified. Occasionally, HSIL extending into glands may be associated with focal epithelial cell necrosis and micronucleoli, without invasion; in such cases, the necrosis is seen associated with the cell group with an otherwise clean background and is not admixed with broken-down blood and inflammation as usually seen in an invasive tumor diathesis[28] (Fig. 5.32).

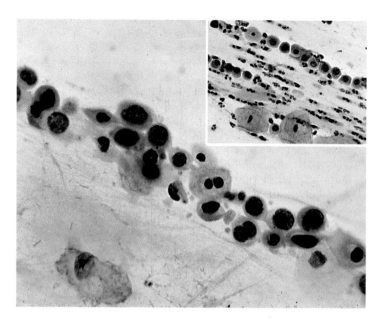

[*] Figure **5.34.** HSIL (CP). At low magnification (*right upper insert*) the pattern of HSIL cells streaming within mucus can mimic histiocytes and endocervical/metaplastic cells. At high power, HSIL can be readily distinguished (see also Figs. 5.35, 4.19, 4.20).

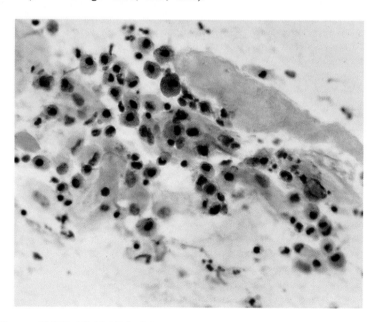

[*] Figure **5.35.** NILM (CP). 34-year-old woman, day 19 of menstrual cycle. Degenerated endocervical cells, seen in a streaming pattern along with thick mucus, is a pattern that has been associated with microglandular hyperplasia. When identified, it is typically from the second half of the menstrual cycle, often in women taking oral contraceptives, and may mimic HSIL at low magnification. Follow-up was NILM.

Management of HSIL

Most women with a cytologic result of HSIL will have biopsy-confirmed CIN 2 or CIN 3 identified at the time of colposcopy.[29] Therefore, the 2001 ASCCP consensus guidelines recommend that if biopsy-confirmed CIN is not identified at colposcopy in a woman with a cytologic interpretation of HSIL, all cytologic and histologic material should be reviewed. If the cytologic interpretation of HSIL is upheld on review, a diagnostic excisional procedure should be performed.[10]

Squamous Cell Carcinoma (Figs. 5.36–5.41)

Definition

A malignant invasive tumor showing differentiation toward squamous cells. The Bethesda System does not subdivide squamous cell carcinoma; however, for descriptive purposes, nonkeratinizing and keratinizing carcinomas are here discussed separately.

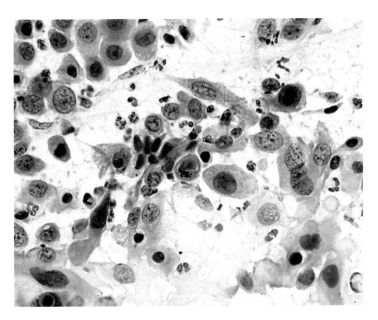

[*] Figure 5.36. Squamous cell carcinoma, keratinizing (CP). Note marked pleomorphism of cell size and shape, cytoplasmic keratinization and tumor diathesis in the background.

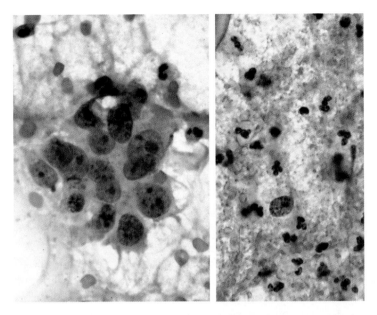

[*] **FIGURE 5.37.** Squamous cell carcinoma (CP). Note tumor diathesis in the background and prominent nucleoli in the malignant cells (*left*). On the right, from a different case, tumor diathesis is prominent and only a naked nucleus is seen in this field (*right*).

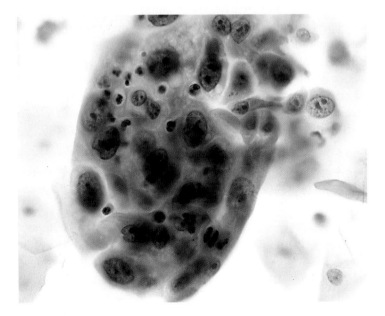

FIGURE 5.38. Squamous cell carcinoma (LBP). Malignant cell clusters tend to show more rounding on LBPs, and distinction between a squamous and glandular lesion may be difficult. Attention should be given to looking for isolated dysplastic cells in the background.

Keratinizing Squamous Cell Carcinoma (Figs. 5.36–5.39)

Criteria

Relatively few cells may be present; often as isolated single cells and less commonly in aggregates.

Marked variation in cellular size and shape is typical, with caudate and spindle cells that frequently contain dense orangeophilic cytoplasm.

Nuclei also vary markedly in size, nuclear membranes may be irregular in configuration, and numerous dense opaque nuclei are often present.

Chromatin pattern, when discernible, is coarsely granular and irregularly distributed with parachromatin clearing.

Macronucleoli may be seen but are less common than in nonkeratinizing squamous cell carcinoma.

Associated keratotic changes ("hyperkeratosis" or "pleomorphic parakeratosis") may be present but are not sufficient for the interpretation of carcinoma in the absence of nuclear abnormalities.

A tumor diathesis may be present, but is usually less than that seen in nonkeratinizing squamous cell carcinomas.

Liquid-Based Preparations

LBPs are often characterized by lower tumor cellularity.[30]

Rounding up of individual cells and cell groups may impart glandular features to squamous tumors, leading to a misinterpretation of adenocarcinoma (Fig. 5.38).

Diathesis is usually identifiable, but can be subtle compared to conventional smears; necrotic material often collects at the periphery of the cell groups, referred to as "clinging diathesis," as opposed to being distributed in the background as in conventional smears[31] (Fig. 5.39).

Nonkeratinizing Squamous Cell Carcinoma (Figs. 5.40, 5.41)

Criteria

Cells occur singly or in syncytial aggregates with poorly defined cell borders.

Cells are frequently somewhat smaller than those of many HSIL, but display most of the features of HSIL.

Nuclei demonstrate markedly irregular distribution of coarsely clumped chromatin.

A tumor diathesis consisting of necrotic debris and old blood is often present.

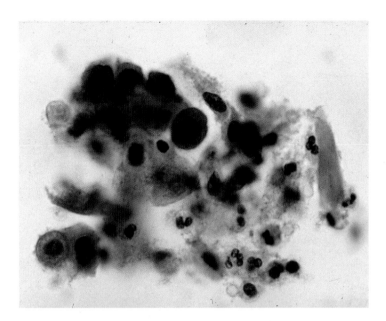

[∗] Figure 5.39. Squamous cell carcinoma, keratinizing (LBP). 68-year-old woman. Diathesis may be more subtle in LBPs and often tends to collect at the periphery of cell groups, a pattern that has been referred to as "clinging diathesis." Follow up was squamous cell carcinoma.

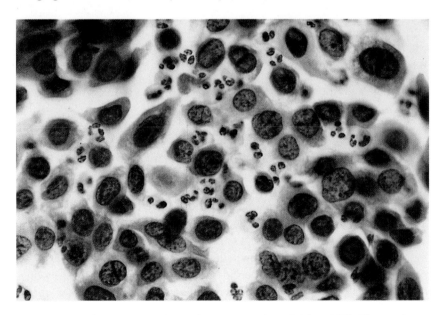

Figure 5.40. Squamous cell carcinoma, non-keratinizing (CP). These dysplastic cells demonstrate nuclear features of HSIL. Pleomorphic cell shapes should raise concern for invasion even though prominent nucleoli and tumor diathesis are absent in this field. Follow-up was squamous cell carcinoma.

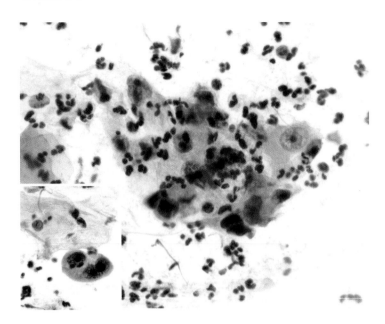

FIGURE 5.41. Squamous cell carcinoma, nonkeratinizing (CP). This less common "large cell variant" shows cells with a moderate amount of cyanophilic cytoplasm and prominent nucleoli.

In addition, large cell variant tumors (Fig. 5.41) may show:

Prominent macronucleoli
Basophilic cytoplasm

Explanatory Notes

Invasive squamous cell carcinoma is the most common malignant neoplasm of the uterine cervix. Previous classifications have divided squamous carcinoma into keratinizing, nonkeratinizing, and small cell types. However, these are often not clearly distinct entities; keratininzing and nonkeratinizing types may coexist on the same slide. A less commonly seen variant of nonkeratinizing squamous cell carcinoma is a "large cell variant" in which the cells may be seen singly and in syncytia and display a moderate amount of cyanophilic cytoplasm and prominent nucleoli (Fig. 5.41).

Tumor diathesis and invasive features may be more difficult to discern in liquid-based preparations resulting in some cancers being interpreted as HSIL.[32]

Historically, "small cell carcinoma" comprised a heterogeneous group of neoplasms, including poorly differentiated squamous cell carcinoma, as well as tumors demonstrating neuroendocrine features (often of the small cell or "oat cell" type). Current classifications limit the use of the term "small cell carcinoma" to non-squamous tumors with evidence of neuroendocrine differentiation. Such tumors, as their counterparts in the lung, are categorized separately in the World Health Organization classification[33] (See Chapter 7, Other Malignant Neoplasms.)

Sample Reports

Example 1:
> Satisfactory for evaluation; endocervical/transformation zone present.
> *Interpretation:*
> Low-grade squamous intraepithelial lesion (LSIL).
> *Note:* Further follow-up as clinically warranted. *(Wright TC Jr, et al. 2001 Consensus Guidelines for the management of women with cervical cytological abnormalities. JAMA 2002;287:2120–2129.)*

Example 2:
> Satisfactory for evaluation.
> *Interpretation:*
> High-grade squamous intraepithelial lesion (HSIL).
> *Note:* Suggest colposcopic examination (with endocervical assessment) as clinically indicated. *(Wright TC Jr, et al. 2001 Consensus Guidelines for the management of women with cervical cytological abnormalities. JAMA 2002;287:2120–2129.)*

Example 3: *Pap test report for a postmenopausal woman*
> Satisfactory for evaluation; endocervical/transformation zone not identified.
> *Interpretation:*
> Low grade squamous intraepithelial lesion arising in an atrophic background.
> *Note:* Suggest colposcopy/biopsy or repeat Pap test 1 week after intravaginal estrogen therapy. *(Wright TC Jr, et al. 2001 Consensus Guidelines for the management of women with cervical cytological abnormalities. JAMA 2002;287:2120–2129.)*

Bethesda System 2001 Workshop Forum Group Moderators:

Thomas C. Wright, M.D., Richard DeMay, M.D., Rose Marie Gatscha, C.T. (ASCP), Lydia Howell, M.D., M.P.H., Ronald D. Luff, M.D., M.P.H., Volker Schneider, M.D., Leo Twiggs, M.D.

References

1. Meisels A, Fortin R, Roy M. Condylomatous lesions of the cervix: II. Cytologic, colposcopic and histopathologic study. *Acta Cytol* 1977;21:379–390.
2. Koss L, Durfee GR. Unusual patterns of squamous epithelium of uterine cervix: cytologic and pathologic study of koilocytotic atypia. *Ann NY Acad Sci* 1956;63: 1245–1261.
3. Bosch FX, Lorincz A, Munoz N, et al. The causal relation between human papillomavirus and cervical cancer. *J Clin Pathol* 2002;55:244–265.
4. Munoz N, Bosch FX, de Sanjose S, et al. Epidemiologic classification of human papillomavirus types associated with cervical cancer. *New Engl J Med* 2003;348: 518–527.
5. Clavel C, Masure M, Bory JP, et al. Hybrid Capture II-based human papillomavirus detection, a sensitive test to detect in routine high-grade cervical lesions: a preliminary study on 1518 women. *Br J Cancer* 1999;80:1306–1311.
6. Ho GY, Bierman R, Beardsley L, et al. Natural history of cervicovaginal papillomavirus infection in young women. *N Engl J Med* 1998;338:423–428.
7. Ylitalo N, Josefsson A, Melbye M, et al. A prospective study showing long-term infection with human papillomavirus 16 before the development of cervical carcinoma in situ. *Cancer Res* 2000;60:6027–6032.
8. Schlecht NF, Kulaga S, Robitaille J, et al. Persistent human papillomavirus infection as a predictor of cervical intraepithelial neoplasia. *JAMA* 2001;286:3106–3114.
9. Ellerbrock TV, Chiasson MA, Bush TJ, et al. Incidence of cervical squamous intraepithelial lesions in HIV-infected women. *JAMA* 2000;283:1031–1037.
10. Wright TC Jr, Cox JT, Massad LS, et al. 2001 Consensus Guidelines for the management of women with cervical cytological abnormalities. *JAMA* 2002;287:2120–2129.
11. Ismail SM, Colelough AB, Dinnen JS, et al. Observer variation in histopathological diagnosis and grading of cervical intraepithelial neoplasia. *Br Med J* 1989;298:707–710.
12. Robertson AJ, Anderson JM, Beck JS, et al. Observer variability in histopathological reporting of cervical biopsy specimens. *J Clin Pathol* 1989;42:231–238.
13. Schneider V. Symposium Part 2: should the Bethesda System Terminology be used in diagnostic surgical pathology? counterpoint. *Int J Gynecol Pathol* 2002;22:13–17.
14. Syrjänen K, Kataja V, Yliskoski M, et al. Natural history of cervical human papillomavirus lesions does not substantiate the biologic relevance of the Bethesda system. *Obstet Gynecol* 1992;79:675–682.
15. Mitchell MF, Tortolero-Luna G, Wright T, et al. Cervical human papillomavirus infection and intraepithelial neoplasia: a review. *J Natl Cancer Inst Monogr* 1996;21: 17–25.

16. Woodhouse SL, Stastny JF, Styer PE, et al. Interobserver variability in subclassification of squamous intraepithelial lesions: Results of the College of American Pathologists Interlaboratory Comparison Program in Cervicovaginal Cytology. *Arch Pathol Lab Med* 1999;123:1079–1084.

17. Cox JT, Solomon D, Schiffman M. Prospective follow-up suggests similar risk of subsequent CIN 2 or 3 among women with CIN 1 or negative colposcopy and directed biopsy. *Am J Obstet Gynecol* 2003;188:1406–1412.

18. Kadish AS, Burk RD, Kress V, et al. Human papillomavirus of different types in precancerous lesions of the uterine cervix: histologic, immunocytochemical and ultrastructural studies. *Hum Pathol* 1986;17:384–392.

19. Willett GD, Kurman RJ, Reid R, et al. Correlation of the histological appearance of intraepithelial neoplasia of the cervix with human papillomavirus types. *Int J Gynecol Pathol* 1989;8:18–25.

20. Wright TC, Ferenczy AF, Kurman RJ. Precancerous lesions of the cervix. In: Kurman RJ, ed. *Blaustein's Pathology of the Female Genital Tract*, 5th Ed. New York: Springer-Verlag, 2002:253–354.

21. The ALTS Group. Human papillomavirus testing for triage of women with cytologic evidence of low-grade squamous intraepithelial lesions: baseline data from a randomized trial. *J Natl Cancer Instit* 2000;92:397–402.

22. Wang SS, Hildesheim A. Viral and host factors in human papillomavirus persistence and progression. *J Natl Cancer Inst Monogr* 2003;31:35–40.

23. Adams KC, Absher KJ, Brill YM, et al. Reproducibility of subclassification of squamous intraepithelial lesions: conventional versus ThinPrep Paps. *J Lower Genital Tract Disease* 2003;7:203–208.

24. Selvaggi SM. Cytologic features of squamous cell carcinoma in situ involving endocervical glands in endocervical cytobrush speicmens. *Acta Cytol* 1994;38:687–692.

25. Selvaggi SM. Cytologic features of high-grade squamous intraepithelial lesions involving endocervical glands on ThinPrep cytology. *Diagn Cytopathol* 2002;26(3): 181–185.

26. Frable WJ. Litigation cells: definition and observations on a cell type in cervical/vaginal smears not addressed by the Bethesda System. *Diagn Cytopathol* 1994;11:213–215.

27. Bosch MM, Rietveld-Scheffers PE, Boon ME. Characteristics of false-negative smears tested in the normal screening situation. *Acta Cytol* 1992;36:711–716.

28. Covell JL, Frierson HF Jr. Intraepithelial neoplasia mimicking microinvasive squamous-cell carcinoma in endocervical brushings. *Diagn Cytopathol* 1992;8(1):18–22.

29. Jones BA, Novis DA. Cervical biopsy-cytology correlation. A College of American Pathologists Q-Probes study of 22, 439 correlations in 348 laboratories. *Arch Pathol Lab Med* 1996;120:523–531.

30. Clark SB, Dawson AE. Invasive squamous-cell carcinoma in ThinPrep specimens: diagnostic clues in the cellular pattern. *Diagn Cytopathol* 2002;26:1–4.

31. Inhorn SL, Wilbur D, Zahniser D, Linder J. Validation of the ThinPrep Papanicolaou test for cervical cancer diagnosis. *J Lower Genital Tract Disease* 1998;2:208–212.

32. Renshaw AA, Young NA, Colgan TJ, et al. Comparison of performance of conventional and ThinPrep gynecologic preparations in the College of American Pathologists gynecologic cytology program. *Arch Pathol Lab Med* 2004;128:17–22.

33. Scully RE, Bonfiglio TA, Kurman RJ, et al. *Histological Typing of Female Genital Tract Tumors*, 2nd Ed. Berlin: Springer-Verlag, 1994.

34. Boot CN, R Massarani-Wafai R, Salhadar A, Weiri J, Wojcik EM. Is LSIL, Cannot Exclude HSIL (LGHSIL) a Valid Pap Test Interpretation? *Mod Pathol* 2005;18(1):61A.

Chapter 6

Epithelial Abnormalities: Glandular

Jamie L. Covell, David C. Wilbur, Barbara Guidos,
Kenneth R. Lee, David C. Chhieng, and Dina R. Mody

Epithelial Cell Abnormalities

Glandular cell

❑ Atypical
  ➢ endocervical cells (NOS or specify in comments)
  ➢ endometrial cells (NOS or specify in comments)
  ➢ glandular cells (NOS or specify in comments)
❑ Atypical
  ➢ endocervical cells, favor neoplastic
  ➢ glandular cells, favor neoplastic
❑ Endocervical adenocarcinoma in situ
❑ Adenocarcinoma
  ➢ endocervical
  ➢ endometrial
  ➢ extrauterine
  ➢ not otherwise specified (NOS)

Background

The 2001 Bethesda System terminology has incorporated changes to the reporting of glandular abnormalities to better reflect current knowledge and understanding of glandular neoplasia in cervical cytology, improve communication among laboratories and clinicians, and thereby facilitate appropriate management of patients.[1] As noted previously, cervical cytology is primarily a screening test for squamous intraepithelial lesions and squamous cell carcinoma; sensitivity for glandular lesions is limited by problems with both sampling and interpretation.[2]

Endocervical adenocarcinoma in situ is considered to be the glandular counterpart of cervical intraepithelial neoplasia (CIN) 3 and the precursor to invasive endocervical adenocarcinoma. Similar human papillomavirus (HPV) types have been demonstrated in most invasive endocervical adeno-carcinomas and adenocarcinoma in situ (AIS).[3,4] Using well-defined criteria, the cytologic interpretation of AIS correlates with histologic outcome. On this basis, AIS is now a separate interpretation in the 2001 Bethesda System. However, no low-grade endocervical glandular entity analogous to low-grade squamous intraepithelial lesion (LSIL) has been identified. A significantly lower rate of detection of HPV DNA in so-called histologic glandular dysplasia suggests that most may be unrelated to cervical carcinogenesis. Therefore, terms such as "endocervical glandular dysplasia" or "low-grade glandular intraepithelial lesion" are not included in Bethesda.[1] The Bethesda intepretation of "Atypical—endocervical, endometrial or glandular—cells" defines an increased level of risk, as opposed to a specific neoplastic precursor entity. Additional highlights of this "atypical" category include:

- The term "atypical glandular cells of undetermined significance" has been eliminated to avoid confusion with the terminology for squamous cell abnormalities (ASC-US).
- Atypical glandular findings should be categorized as to cell type of origin (endocervical or endometrial) whenever possible, as the clinical workup and management for patients with glandular abnormalities may vary significantly depending upon the cell type; otherwise, the generic "Atypical glandular cells" (AGC) terminology is used.
- "Atypical endocervical cells" and "atypical glandular cells" may be further qualified as "Favor neoplastic." The qualifier "Favor reactive" was considered to be potentially misleading and, therefore, was eliminated. If not further qualified, "Not otherwise specified" (NOS) may be used.
- "Atypical endometrial cells" are not further qualified, reflecting the difficulty of reliably further subclassifying this category.

Atypical Glandular Cells

Atypical Endocervical Cells

Definition

Endocervical-type cells display nuclear atypia that exceeds obvious reactive or reparative changes but that lack unequivocal features of endocervical adenocarcinoma in situ or invasive adenocarcinoma.

Atypical Endocervical Cells: NOS (Figs. 6.1–6.6)

Criteria

Cells occur in sheets and strips with some cell crowding and nuclear overlap.

Nuclear enlargement, up to three to five times the area of normal endocervical nuclei, may be seen.

Some variation in nuclear size and shape is present.

Mild hyperchromasia is frequently evident.

Nucleoli may be present.

Mitotic figures are rare.

Cytoplasm may be fairly abundant, but the nuclear/cytoplasmic (N/C) ratio is increased.

Distinct cell borders often are discernible.

Liquid-Based Preparations

Groups are more rounded and three dimensional with piled-up layers of cells, making individual cells in the center difficult to visualize.

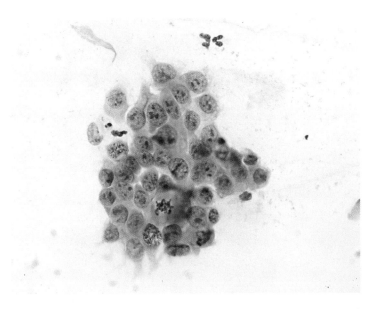

[∗] FIGURE **6.1.** Atypical endocervical cells, most likely from a reparative process (conventional preparation, CP). Routine screen from a 39-year-old woman. Sheet of cells that demonstrate nuclear enlargement, increased nuclear/cytoplasmic (N/C) ratios, prominent, sometimes multiple nucleoli, and mitotic activity. Three year follow-up was NILM. [[∗] Bethesda Interobserver Reproducibility Project (BIRP) image (see xviii Introduction).]

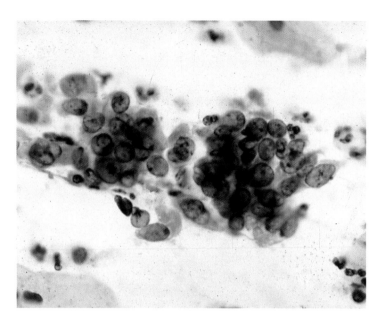

FIGURE 6.2. Atypical endocervical cells, not otherwise specified (NOS) (CP). Cells are characterized by round or oval nuclei with nuclear enlargement, disordered arrangement, and occasional nucleoli. Follow-up was endocervical adenocarcinoma in situ (AIS).

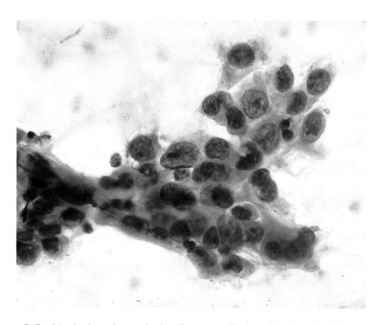

FIGURE 6.3. Atypical endocervical cells, most likely related to ionizing radiation therapy (CP). 54-year-old woman 4 months status post radiation therapy for cervical carcinoma. Sheet of cells which show nuclear enlargement, marked variation in nuclear size, prominent nucleoli, and distinct cell borders. Follow-up was NILM.

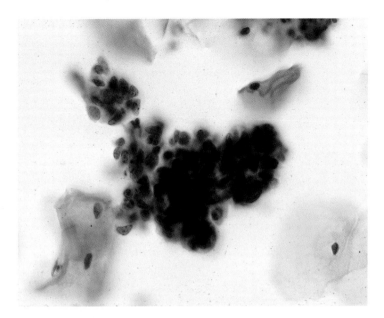

FIGURE 6.4. Atypical endocervical cells, NOS (liquid-based preparation, LBP). The cells are often seen in three-dimensional clusters that can make visualization of cells in the center difficult.

FIGURE 6.5. Atypical glandular cells (LBP). 45-year-old woman with an intrauterine device (IUD). The presumed endocervical cells demonstrate nuclear enlargement, nucleoli, and cytoplasmic vacuolization, consistent with changes associated with presence of an IUD. In the absence of a clinical history of IUD, such changes may be considered atypical.

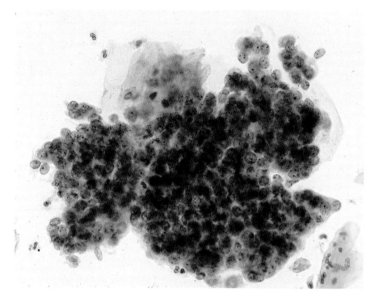

[*] Figure 6.6. Case reported as "atypical endocervical cells," NOS (CP). Cluster of cells shows crowding and overlapping of nuclei, nuclear enlargement, chromocenters, and small nucleoli. Follow-up biopsies showed high-grade cervical intraepithelial neoplasia (CIN).

Atypical Endocervical Cells, Favor Neoplastic (Figs. 6.7, 6.8)

Definition

Cell morphology, either quantitatively or qualitatively, falls just short of an interpretation of endocervical adenocarcinoma in situ or invasive adenocarcinoma.

Criteria

Abnormal cells occur in sheets and strips with nuclear crowding and overlap.

Rare cell groups may show rosetting or feathering.

Nuclei are enlarged with some hyperchromasia.

Occasional mitoses may be seen.

Nuclear/cytoplasmic ratios are increased, quantity of cytoplasm is diminished, and cell borders may be ill defined.

Liquid-Based Preparations

Groups may be three dimensional, thick, with layers of cells obscuring central nuclear detail.

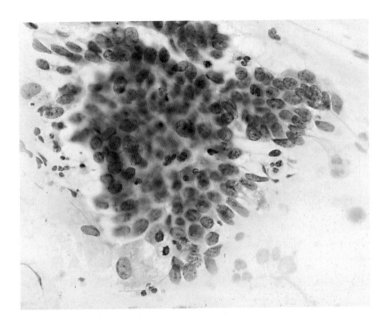

FIGURE 6.7. Atypical endocervical cells, favor neoplastic (CP). Routine screen from a 29-year-old woman. Sheet of crowded cells with increased N/C ratios and mitotic activity. Note feathering at the edges of the sheet. Follow-up was endocervical AIS.

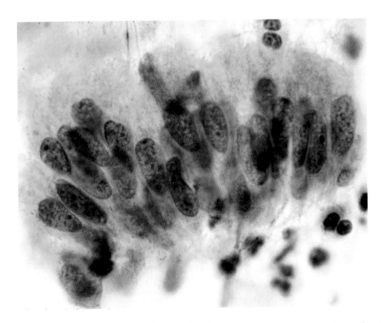

FIGURE 6.8. Atypical endocervical cells, favor neoplastic (CP). Pseudostratified strip of endocervical cells with enlarged, elongated nuclei and some chromatin granularity.

Atypical Endometrial Cells (Figs. 6.9–6.12)

Criteria

Cells occur in small groups, usually 5 to 10 cells per group.
Nuclei are slightly enlarged compared to normal endometrial cells.
Mild hyperchromasia may be seen.
Small nucleoli may be present.
Scant cytoplasm is occasionally vacuolated.
Cell borders are ill defined.

Liquid-Based Preparations

Nuclear hyperchromasia may be more prominent.
Nucleoli may be more prominent.
Note: Benign exfoliated (shed/ menstrual) endometrial cells may show greater nuclear pleomorphism than is seen in conventional smears (see Chapter 3, Endometrial Cells). These changes are likely due to improved visualization of degenerating endometrial cells resulting from clearing of blood, inflammation, and debris in menstrual liquid-based preparations (LBPs) and should not be overinterpreted as "atypical."

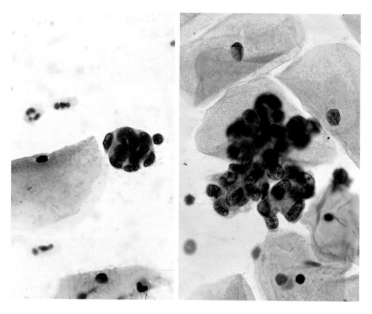

Figure 6.9. Atypical endometrial cells (CP). 82-year-old woman with post-menopausal bleeding. Three-dimensional groups of small cells with mildly hyperchromatic nuclei, small nucleoli, and occasionally vacuolated cytoplasm. Follow up was endometrial hyperplasia.

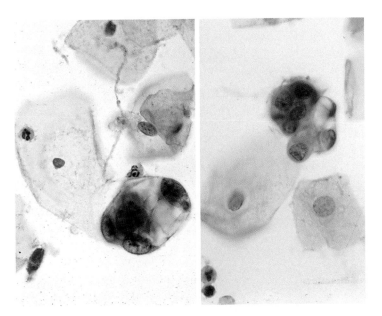

[*] FIGURE 6.10. Atypical endometrial cells (LBP). Small groups of cells with slightly enlarged nuclei, small nucleoli, and vacuolated cytoplasm. Left: 63-year-old woman. Follow up endometrial adenocarcinoma grade 1. Right: 55-year-old woman. Follow up was endometrial hyperplasia.

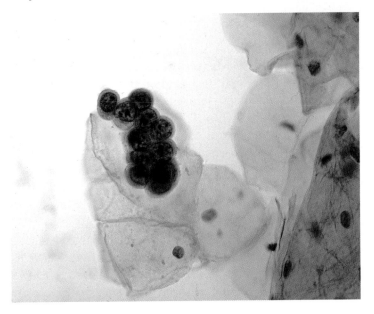

[*] FIGURE 6.11. Atypical endometrial cells (LBP). 63-year-old woman with postmenopausal bleeding. Aggregate of small cells with slightly enlarged round or oval nuclei, small nucleoli, and finely vacuolated cytoplasm. Follow up was endometrial adenocarcinoma grade 1.

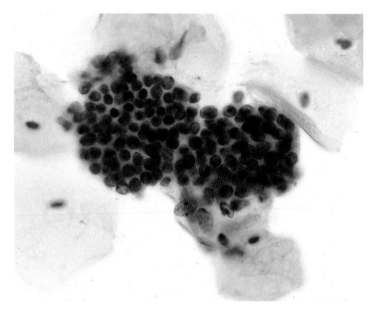

[*] FIGURE 6.12. Atypical endometrial cells (LBP). 52-year-old woman on hormone replacement therapy. Three-dimensional grouping of small cells with crowded round or oval nuclei. Follow up was endometrial hyperplasia.

Explanatory Notes

The interpretation of "atypical glandular cells" (AGC) should be qualified, if possible, to indicate whether the cells are thought to be of endocervical or endometrial origin. If the origin of the cells cannot be determined, the generic "glandular" term is used. Atypical endocervical cells should be further qualified when a particular entity, including neoplasia, is favored. Atypical endometrial cells are generally not further qualified as favor neoplastic since this is a difficult and poorly reproducible distinction. However, specific comments can be appended if clinical findings/history are available (e.g. presence of IUD, polyp).[1]

Endocervical and endometrial glandular cells may show a variety of cellular changes associated with various benign processes in the endocervical canal and endometrium.[5] Many of these reactive changes are not specific for any particular disease entity, but are of significance as mimics of glandular neoplasia in cervical cytology.[6] Reactive endocervical cells are characterized by the presence of a honeycomb or sheetlike arrangement with abundant cytoplasm, well-defined cell borders, and minimal nuclear overlap. Some degree of pleomorphism of cell size and nuclear enlargement may be noted; however, the nuclei remain round or oval with smooth contours and bland chromatin. Nucleoli may be prominent and multinucleation can occur, especially in cases of repair and inflammation. Cytoplasmic mucin may be diminished, giving the cell cluster a more hyperchromatic appearance. This constellation of reactive changes should be considered as "Negative for intraepithelial lesion or malignancy" and not included in the AGC category (see Figs. 2.17, 2.18, 2.22).[1]

"Atypical endocervical cells" may be used for cases demonstrating some, but not all, of the criteria necessary for endocervical adenocarcinoma in situ (AIS) or invasive adenocarcinoma. These features include nuclear enlargement, crowding, variation in size, and hyperchromasia. Some non-neoplastic processes that may show atypical cellular changes and lead to interpretive difficulty include lower uterine segment sampling, tubal metaplasia, repair, endocervical polyps, microglandular hyperplasia, Aria–Stella change, and effects of ionizing radiation[5,7–10] (see Fig. 6.3).

Vigorous sampling using an endocervical brush may transfer large hyperchromatic groups of intact normal endocervical cells to the slide, resulting in so-called brush artifact. Such hyperchromatic groups may cause concern due to the inability to visualize centrally placed cells. These groups should be carefully evaluated for nuclear and architectural features of glandular or squamous neoplasia before rendering an "atypical" interpretation.

Tubal metaplasia is usually categorized as "Negative for intraepithelial lesion or malignancy" (NILM). However, it is also a significant pitfall in the interpretation of glandular changes.[6–8,10] Only if the findings are sufficiently atypical to raise concern for neoplasia should the interpretation "atypical endocervical cells" be used. The nuclei of cells from tubal metaplasia are often enlarged, hyperchromatic, and pseudostratified, resembling those seen in endocervical adenocarcinoma in situ (AIS) (Figs. 6.13–6.15). Although some architectural and cytologic features overlap with AIS, the nuclei in tubal metaplasia tend to be round or oval and display more finely granular, evenly dispersed chromatin. Feathered edges, rosette formation, and mitoses may be seen, but they are less common compared to classic AIS. The most helpful criterion is the presence of cilia that may require high-powered microscopic evaluation of cell clusters to be appreciated. Although the presence of rare ciliated abnormal cells has been described in glandular neoplasia, terminal bars and cilia are commonly seen in tubal metaplasia. In addition, intermixed goblet cells and slender "peg" cells may be identified (Figs. 6.16, 2.35).

The distinction of cytologically benign versus atypical endometrial cells is based primarily on the criterion of increased nuclear size. The presence of atypical endometrial cells, like their cytologically bland counterparts, may be associated with the presence of endometrial polyps, chronic endometritis, an intrauterine device (IUD), endometrial hyperplasia, or endometrial carcinoma (see Fig. 6.10). Caution should be used in the interpretation of atypia in endometrial material on liquid-based preparations because shed/menstrual endometrial cells may show significantly greater

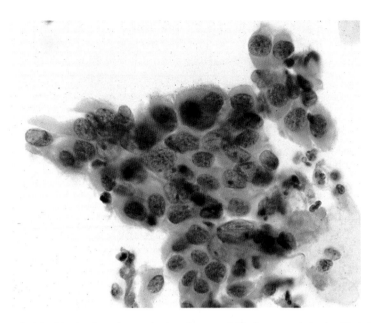

FIGURE 6.13. Atypical endocervical cells, most likely associated with tubal metaplasia (CP). Routine screen from a 38-year-old woman. Sheet of cells having enlarged, variably sized nuclei with some nuclear crowding and overlap. Note cilia at upper edge of sheet. Follow up was tubal metaplasia on biopsy.

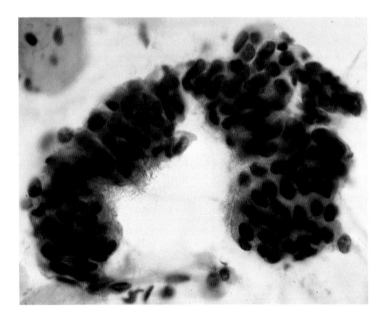

FIGURE 6.14. Atypical endocervical cells, most likely associated with tubal metaplasia (CP). Three-dimensional group of cells with pseudostratification and crowding of cells. Note the presence of cilia on some cells.

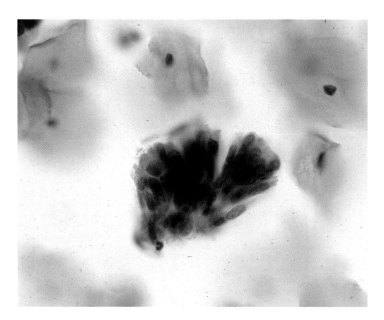

FIGURE 6.15. Atypical endocervical cells, most likely associated with tubal metaplasia (LBP). Cell group demonstrates crowding, pseudostratification, and oval or elongated nuclei. Note the presence of cilia on some cells.

pleomorphism of nuclear size and shape than is seen in conventional smears (see Figs. 3.2, 3.4). Clinical information can be helpful in avoiding an overinterpretation in these cases.[5,11]

High-grade squamous intraepithelial lesion (HSIL) involving gland spaces may present as contoured clusters mimicking the appearance of a glandular lesion (see Fig. 6.6). Cell groups are composed of tightly packed cells with high nuclear/cytoplasmic ratios and hyperchromatic nuclei with coarsely granular chromatin. The cytoplasm often has no specific differentiation. Flattening of cells at the periphery of the cluster, loss of cell polarity within the clusters, and the presence of isolated dysplastic squamous cells in the background can be very helpful features to suggest HSIL (see Figs. 5.13, 5.14, 5.29, 5.30, 5.31). HSIL involving gland spaces also lacks specific architectural features of AIS such as feathering, rosettes, pseudostratified strips of columnar cells, and maintenance of cell polarity (see Figs. 6.19, 6.20).[5,12–14]

Management of AGC

Consensus guidelines from the American Society for Colposcopy and Cervical Pathology (ASCCP) include recommendations for the initial workup and subsequent management of women with glandular abnormalities based on 2001 Bethesda terminology.[15] Follow-up of AGC cytologic in-

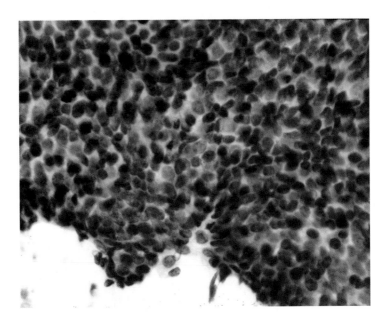

FIGURE 6.16. Atypical endocervical cells, probably derived from tubal metaplasia (CP). Cell groups from tubal metaplasia may raise the differential diagnosis of endocervical adenocarcinoma in situ (AIS). It is useful to note that due to the presence of mucin in goblet cells overlying some nuclei, and the variety of cell types (goblet, ciliated, and peg) in tubal metaplasia, scattered nuclei demonstrate relative hypochromasia or a "washed-out" appearance and lack the monotony of changes characteristic of AIS (contrast with Fig. 6.17).

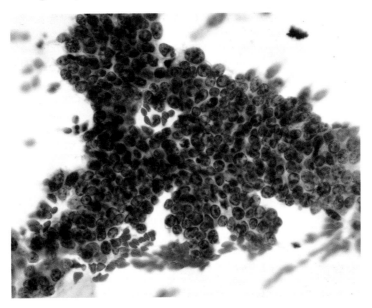

[*] **FIGURE 6.17.** Endocervical adenocarcinoma in situ (CP). Sheet of crowded cells with enlarged, hyperchromatic nuclei, increased nuclear/cytoplasmic ratios, and feathering at the periphery of the sheet. Note the monotony of the hyperchromatic nuclei as contrasted with the more variable nuclear changes in tubal metaplasia (see Fig. 6.16).

terpretations shows that high-grade lesions are identified in 10% to 40% of cases and are more often squamous (CIN 2 or 3) than glandular[5,12] (Fig. 5.33 and 6.6).

Endocervical Adenocarcinoma In Situ (AIS) (Figs. 6.17–6.24)

Definition

High-grade endocervical glandular lesion that is characterized by nuclear enlargement, hyperchromasia, stratification, and mitotic activity, but without invasion.

Criteria

Cells occur in sheets, clusters, strips, and rosettes with nuclear crowding and overlap, and loss of honeycomb pattern. Single abnormal cells are uncommon.

Some cells show a definite columnar appearance.

Cell clusters have a palisading nuclear arrangement with nuclei and cytoplasmic tags protruding from the periphery ("feathering").

Nuclei are enlarged, variably sized, oval or elongated in shape, and stratified.

Nuclear hyperchromasia with evenly dispersed, coarsely granular chromatin is characteristic.

Nucleoli are usually small or inconspicuous.

Mitoses and apoptotic bodies are commonly seen.

Nuclear/cytoplasmic ratios are increased; the quantity of cytoplasm, as well as mucin, is diminished.

Background is typically clean (no tumor diathesis or inflammatory debris).

Abnormal squamous cells may be present if there is a coexisting squamous lesion.

Liquid-Based Preparations

Cellularity is variable; occasionally abundant abnormal cells are present. Single intact cells are more easily found.

Hyperchromatic crowded groups are smaller, denser, and more three dimensional with smoother, sharper margins.

Pseudostratified strips of cells, often presenting as short "bird tail"-like arrangements, may be the most prominent architectural feature (Fig. 6.23).

Architectural features of peripheral feathering, rosettes, and cell strips have a more subtle presentation.

Nuclear chromatin may be coarse or finely granular, often with a more open appearance.

Nucleoli may be more readily visible (Fig. 6.24).

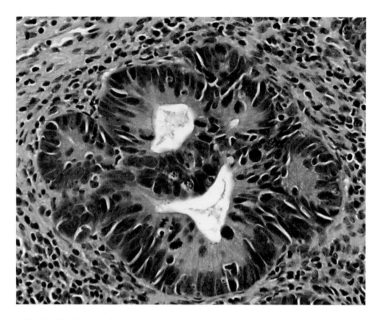

FIGURE 6.18. Endocervical adenocarcinoma in situ (histology, H&E).

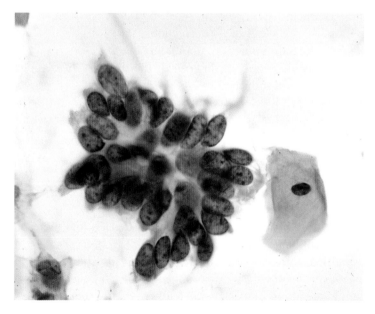

FIGURE 6.19. Endocervical adenocarcinoma in situ (CP). The typically oval nuclei are crowded with nuclear overlapping and show hyperchromasia with evenly distributed but coarsely granular chromatin.

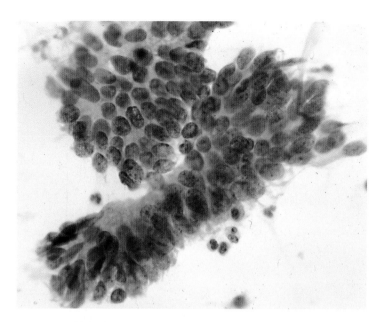

FIGURE 6.20. Endocervical adenocarcinoma in situ (CP). Pseudostratified strip of cells demonstrating crowding, nuclear enlargement, and peripheral feathering.

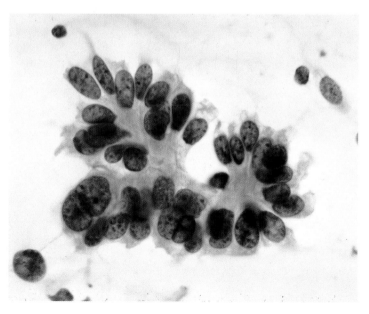

[*] **FIGURE 6.21.** Endocervical adenocarcinoma in situ (CP). Cell group in a rosette-like arrangement. Nuclei are oval or elongated, hyperchromatic, and have granular, evenly distributed chromatin.

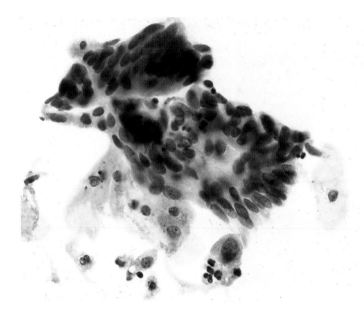

[*] **FIGURE 6.22.** Endocervical adenocarcinoma in situ (LBP). 64-year-old woman with prior abnormal cytology. Cell groups in LBPs may be more three-dimensional with sharper, smoother margins, and feathering may have a more subtle presentation. Follow-up was AIS with small focus of invasion.

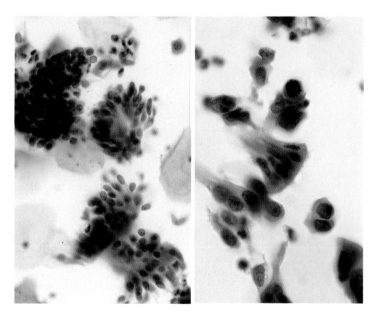

[*] **FIGURE 6.23.** Endocervical adenocarcinoma in situ (LBP). Routine screen from a 25-year-old woman. Pseudostratified strips of cells often present as short "birdtail like" arrangements in LBPs as seen on the *right side* of this image. Feathering, although less prominent than in conventional smears, is demonstrated on the *left*. Follow-up was AIS.

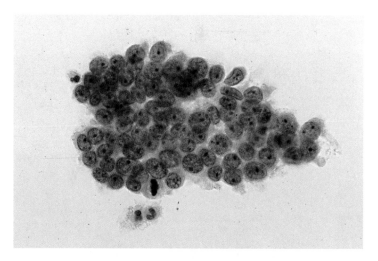

FIGURE 6.24. Endocervial adenocarcinoma in situ (LBP). AIS may occasionally demonstrate nucleoli, raising the differential of invasive endocervical carcinoma (see Fig. 6.30).

Endocervical Adenocarcinoma (Figs. 6.25–6.31)

Cytologic criteria overlap those outlined for AIS, but may show features of invasion.

Criteria

Abundant abnormal cells, typically with columnar configuration.

Single cells, two-dimensional sheets, or three-dimensional clusters and syncytial aggregates are commonly seen.

Enlarged, pleomorphic nuclei demonstrate irregular chromatin distribution, parachromatin clearing, and nuclear membrane irregularities.

Macronucleoli may be present.

Cytoplasm is usually finely vacuolated.

Necrotic tumor diathesis may be seen.

Abnormal squamous cells may additionally be present, representing a co-existing squamous lesion or the squamous component of an adenocarcinoma showing partial squamous differentiation.

Liquid-Based Preparations

Three-dimensional clusters are more common.

Chromatin is more open (vesicular) with irregular chromatin distribution and parachromatin clearing.

Tumor diathesis may be less prominent and seen as debris clinging to the periphery of clusters of abnormal cells or as coagulated debris (Fig. 6.30).

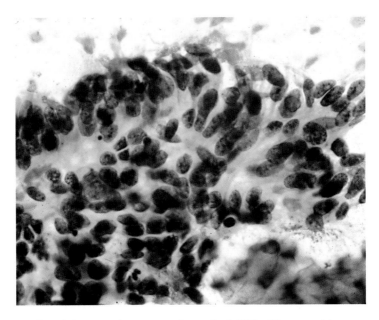

FIGURE 6.25. Adenocarcinoma, endocervical (CP). 32-year-old woman with abnormal cervix on exam, menstrual smear. Cytologic features may overlap with those of endocervical adenocarcinoma in situ. Follow up was invasive endocervical adenocarcinoma.

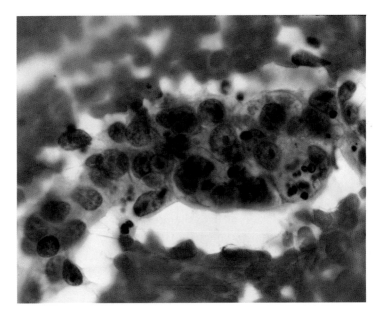

FIGURE 6.26. Adenocarcinoma, endocervical (CP). Nuclei are enlarged and pleomorphic with irregular chromatin distribution and prominent or macronucleoli. Cytoplasm is finely vacuolated.

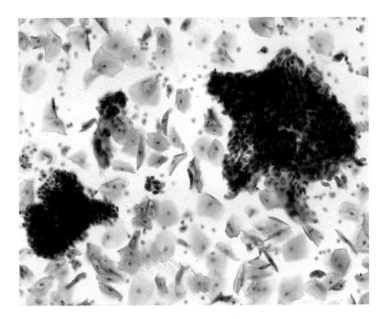

FIGURE 6.27. Adenocarcinoma, endocervical (LBP). Large cell groups may be thick and three-dimensional, making architecture more difficult to interpret and visualization of cell nuclei more problematic.

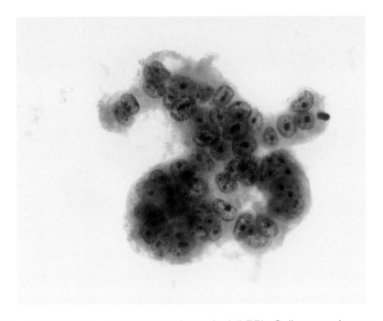

FIGURE 6.28. Adenocarcinoma, endocervical (LBP). Cell group demonstrates glandular architecture and large overlapping nuclei, irregular chromatin distribution, and nucleoli.

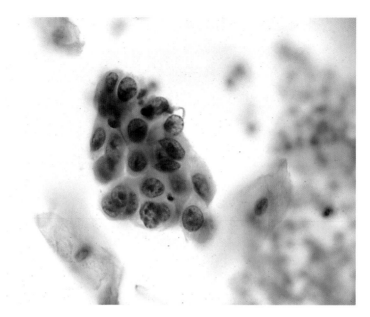

[*] **FIGURE 6.29.** Adenocarcinoma, endocervical (LBP). 46-year-old woman. Cell nuclei may have more open vesicular chromatin with irregular distribution and parachromatin clearing as well as macronucleoli. Follow up was invasive endocervical adenocarcinoma.

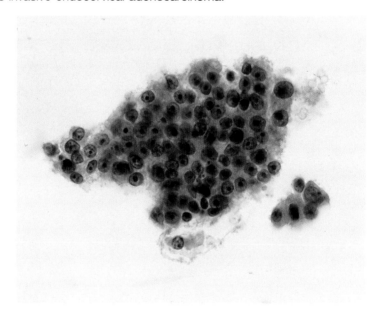

[*] **FIGURE 6.30.** Adenocarcinoma, endocervical (LBP). 39-year-old woman, day 12 of menstrual cycle. Tumor diathesis may be less prominent and seen as debris clinging to the periphery of the abnormal cell clusters in LBPs. Follow up was invasive endocervical adenocarcinoma.

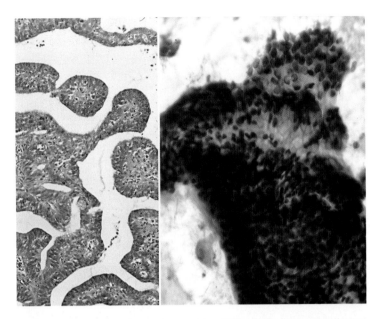

FIGURE 6.31. Adenocarcinoma, villoglandular (CP). A rare neoplasm of the cervix, villoglandular carcinoma may demonstrate large cohesive groups of endocervical cells with nuclear crowding and loss of normal honeycomb pattern, with true papillary clusters being characteristic. Cytologic atypia is often minimal, emphasizing the importance of appreciating the low-power architectural abnormalities of this neoplasm. (*Right*, cytology (CP); *left*, histology (H&E).)

Explanatory Notes

The cytologic interpretation of endocervical adenocarcinoma in situ can be difficult and should only be made in cases where sufficient criteria are present. In problematic cases, the interpretation of "atypical endocervical/glandular cells, favor neoplastic" is justified.[1] Criteria described for AIS are the features for the most common endocervical form.[5,10,12,16–18] Although uncommon, variant forms of AIS exist (e.g., intestinal, endometrioid, clear cell) that may show other morphologic features[5,12,19,20] (Fig. 6.32).

The cytologic presentations of various histologic types of invasive endocervical adenocarcinoma have been described (Fig. 6.31).[5,12,18] An invasive adenocarcinoma should be strongly considered in the presence of tumor diathesis, nuclear clearing with uneven distribution of chromatin, or macro-nucleoli.[6] However, in some well-differentiated cases, tumor diathesis and macronucleoli may be absent.

Glandular lesions demonstrate several morphologic differences in liquid-based preparations in comparison to conventional smears.[18] Cell groups tend to be thick, spherical, and three-dimensional, making archi-

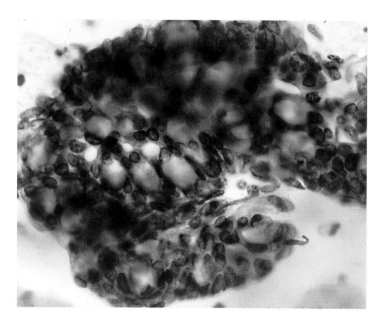

FIGURE 6.32. Endocervical adenocarcinoma in situ, intestinal type (CP). Cells show nuclear crowding and overlap and have elongated nuclei. Note numerous goblet-type cells.

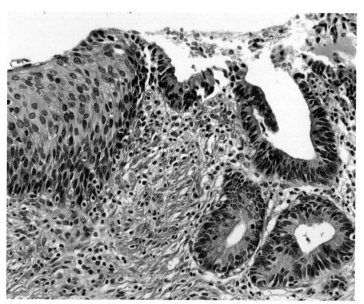

FIGURE 6.33. Glandular and squamous lesions may coexist. Cervical intraepithelial neoplasia (CIN) grade 2 is seen on the squamous epithelial surface on the *left side* of this image, and endocervical adenocarcinoma in situ is present in gland spaces on the *right* (histology, H&E). (© 2001 American Society for Clinical Pathology. Reprinted with permission.)

tecture more difficult to interpret and visualization of the cell nuclei more problematic. In fact, nuclei within the central portions of groups may be completely obscured (see Fig. 6.27). Close and careful scrutiny is essential to correctly categorize these clusters as glandular in origin. Isolated abnormal cells are more frequently seen. Because of the more open chromatin patterns seen in liquid-based preparations, nucleoli are more prominent. However, tumor diathesis may be less apparent, consisting of aggregates of proteinaceous and inflammatory debris often found clinging to the surface of individual cells or cell clusters in a pattern that has been referred to as "clinging diathesis" (see Fig. 6.30).

The possibility of coexisting glandular and squamous lesions in the cervix should always be considered when making an interpretation of endocervical AIS (Fig. 6.33).[5] In some studies, up to half of AIS lesions have a coexisting squamous intraepithelial lesion, usually of high grade. Often, the cytoplasmic features and the cell arrangements differentiate the two neoplastic processes.

Endometrial Adenocarcinoma (Figs. 6.34–6.38)

Criteria

Cells typically occur singly or in small, tight clusters.

In well-differentiated tumors, nuclei may be only slightly enlarged compared to non-neoplastic endometrial cells, becoming larger with increasing grade of the tumor.

Variation in nuclear size and loss of nuclear polarity are evident.

Nuclei display moderate hyperchromasia, irregular chromatin distribution, and parachromatin clearing, particularly in high-grade tumors.

Small to prominent nucleoli are present; nucleoli become larger with increasing grade of tumor.

Cytoplasm is typically scant, cyanophilic, and often vacuolated; intracytoplasmic neutrophils are common.

A finely granular or "watery" tumor diathesis is variably present.

Liquid-Based Preparation

Three-dimensional groups and clusters or papillary configurations may be present.

Nuclei tend to be larger with more open chromatin.

Tumor diathesis may be less prominent, and seen as finely granular debris clinging to the periphery of clusters of abnormal cells or as coagulated debris.

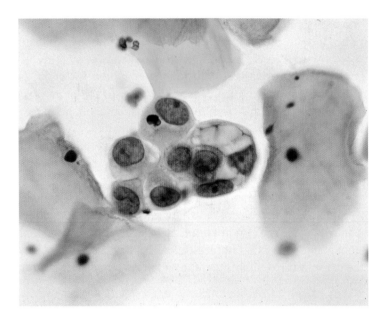

FIGURE 6.34. Adenocarcinoma, endometrial, low grade (CP). Small cluster of cells with slightly enlarged nuclei, small nucleoli, and vacuolated cytoplasm.

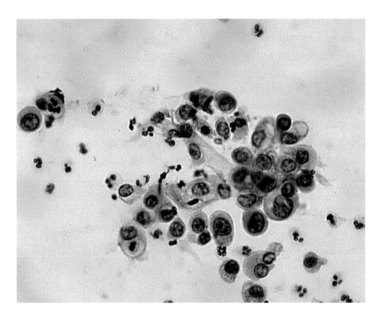

[*] **FIGURE 6.35.** Adenocarcinoma, endometrial, low grade (CP). 61-year-old woman with postmenopausal bleeding. Loose aggregate of small epithelial cells with slightly enlarged nuclei, small nucleoli, and vacuolated cytoplasm. Note watery diathesis in background. Follow up was endometrial adenocarcinoma grade 1.

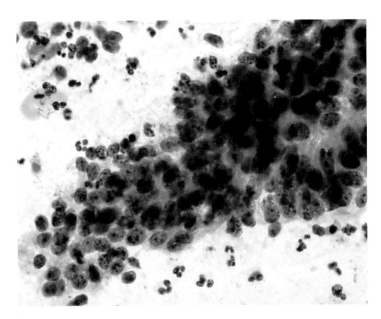

FIGURE 6.36. Adenocarcinoma, endometrial, high grade (CP). 58-year-old woman with postmenopausal bleeding. Nuclei in higher-grade tumors are larger and display moderate hyperchromasia with irregular chromatin distribution. Note finely granular diathesis in background. Follow up was high grade endometrial adenocarcinoma.

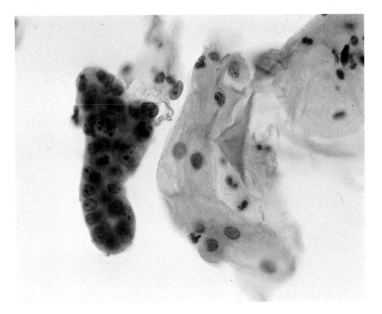

[*] FIGURE 6.37. Adenocarcinoma, endometrial (LBP). 67-year-old woman with postmenopausal bleeding. Three-dimensional cell groups and papillary clusters may be seen. Follow up was endometrial adenocarcinoma grade 1–2.

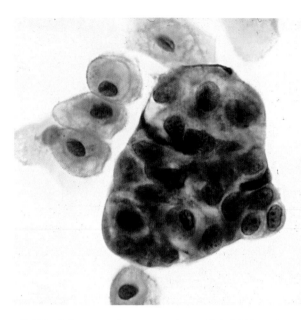

FIGURE 6.38. Adenocarcinoma, endometrial (LBP). 64-year-old woman. Papillary serous carcinomas may resemble their ovarian counterparts and present with papillary groups, large cell size, and prominent nucleoli. Follow up was papillary serous adenocarcinoma of endometrium.

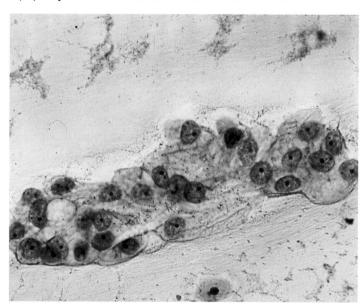

[*] **FIGURE 6.39.** Adenocarcinoma, endometrial, high grade (CP). Tumor diathesis, if present, is watery and may be difficult to appreciate. (Reprinted with permission from Kurman RJ (ed.), *Blaustein's Pathology of the Female Genital Tract*, Fourth Edition, Springer-Verlag, New York, © 1994.)

Explanatory Notes

The cytologic findings in endometrial adenocarcinoma are largely dependent upon the grade of the tumor. Grade 1 tumors tend to shed few abnormal cells with minimal cytologic atypia and would typically be interpreted as atypical endometrial cells (see Fig. 6.10). Cytologic detection of endometrial adenocarcinoma, especially well-differentiated tumors, in cervical specimens is limited by the small number of well-preserved abnormal cells and the subtlety of their cellular alterations. In contrast to endocervical adenocarcinomas that are directly sampled, the detection of endometrial carcinomas in cervical cytology depends on exfoliated cells being present in the collected specimen. Thus, there are generally fewer abnormal cells present as compared to endocervical cancers (see Figs. 6.34, 6.37). In addition, the malignant cells from endometrial carcinomas generally have a smaller cell and nuclear size, nucleoli are less prominent, and tumor diathesis if present is "watery" and more difficult to appreciate[5,6,11,12] (Fig. 6.35). High-grade endometrial serous carcinomas morphologically resemble their ovarian counterpart with papillary fragments, large cell size, and prominent nucleoli (see Fig. 6.38).

Extrauterine Adenocarcinoma (Figs. 6.40–6.42)

When cells diagnostic of adenocarcinoma occur in association with a clean background or with morphology unusual for tumors of the uterus or cervix, an extrauterine neoplasm should be considered. Sources in the female genital tract include the ovary and, less commonly, the fallopian tube.[6,12] Although not specific, the presence of papillary clusters and psammoma bodies suggests an ovarian carcinoma (Fig. 6.40). Because they are exfoliated and travel from distant sites, the malignant cells may show degenerative changes. The lack of tumor diathesis is a clue to the noncervical origin of the malignant cells. When diathesis is present, it is usually associated with metastasis to the uterus or vagina.[12] Other tumors metastatic to the cervix or uterus are considered in Chapter 7, Other Malignant Neoplasms.

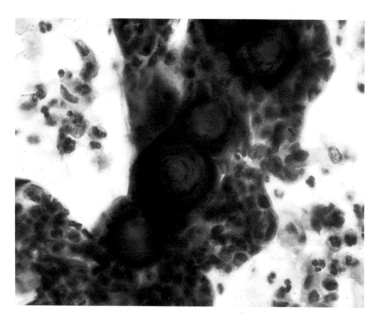

Figure 6.40. Adenocarcinoma, extrauterine (CP). 70-year-old woman with large pelvic mass and ascites. Ovarian carcinoma may be characterized by papillary configurations and psammomatous calcifications (psammoma bodies). Follow up showed an ovarian primary.

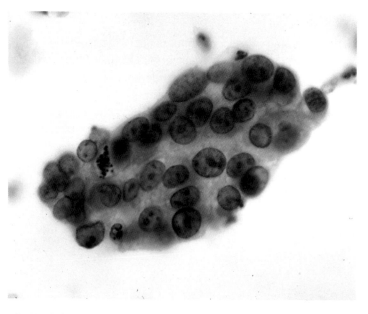

Figure 6.41. Adenocarcinoma, extrauterine (CP). Clusters of cells from ovarian carcinoma have enlarged, variably-sized round or oval nuclei with prominent macronucleoli. The background is typically clean.

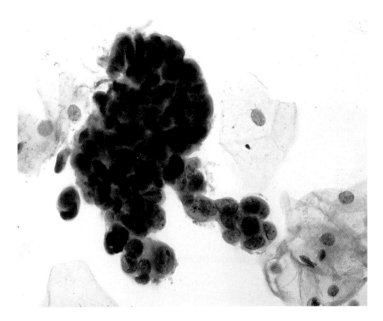

FIGURE 6.42. Adenocarcinoma, extrauterine (LBP). 66-year-old woman with pelvic mass and ascites. Papillary clusters from ovarian carcinoma may be three dimensional, making evaluation of the component cells difficult. Follow up showed intraabdominal dissemination of ovarian carcinoma.

Sample Reports

Example 1:
  Satisfactory for evaluation; endocervical/transformation zone present.
  Interpretation:
  Atypical endometrial cells present (not otherwise specified).

Example 2:
  Satisfactory for evaluation; endocervical/transformation zone present.
  Interpretation:
  Atypical endocervical cells present, mostly likely derived from tubal
    metaplasia.

Example 3:
  Satisfactory for evaluation; endocervical/transformation zone present.
  Interpretation:
  Atypical glandular cells present, favor neoplastic.
  Note: Suggest colposcopy (with endocervical sampling) and endome-
    trial sampling (if >35 years old or abnormal bleeding) as clinically
    indicated. *(Wright TC Jr, Cox JT, Massad LS, Twiggs LB, Wilkin-
    son EJ. 2001 Consensus Guidelines for the management of women with
    cervical cytological abnormalities. JAMA 2002;287:2120–2129.)*

Example 4:
  Satisfactory for evaluation.
  Interpretation:
  Endocervical adenocarcinoma in situ.

Example 5:
  Satisfactory for evaluation.
  Interpretation:
  Adenocarcinoma, favor endometrial origin.

Bethesda System 2001 Workshop Forum Group Moderators:

David Wilbur, M.D., David Chhieng, M.D., J. Thomas Cox, M.D., Jamie
Covell, B.S., C.T. (ASCP), Barbara Guidos, S.C.T. (ASCP), Kenneth Lee,
M.D., Dina Mody, M.D.

References

1. Solomon D, Davey D, Kurman R, et al. The 2001 Bethesda System: terminology for reporting results of cervical cytology. *JAMA* 2002;287:2114–2119.
2. Moriarty AT, Wilbur DC. Those gland problems in cervical cytology: faith or fact? Observations from the Bethesda 2001 terminology conference. *Diagn Cytopathol* 2003;28:171–174.
3. Ronnett BM, Manos MM, Ransley JE, et al. Atypical glandular cells of undetermined significance (AGUS): cytopathologic features, histopathologic results, and human papillomavirus DNA detection. *Hum Pathol* 1999;30:816–825.
4. Pirog EC, Kleter B, Olgac S, et al. Prevalence of human papillomavirus DNA in different histologic subtypes of cervical adenocarcinoma. *Am J Pathol* 2000;157: 1055–1062.
5. Wilbur DC. The cytology of the endocervix, endometrium, and upper female genital tract. In: Bonfiglio TA, Erozan YS, eds. *Gynecologic Cytopathology*. Philadelphia: Lippincott-Raven, 1997:107–156.
6. Kurman RJ, Solomon D. *The Bethesda System for Reporting Cervical/Vaginal Cytologic Diagnoses*. New York: Springer-Verlag, 1994:64–79.
7. Babkowski RC, Wilbur DC, Rutowski M, et al. The effects of endocervical canal topography, tubal metaplasia, and high canal sampling on the cytologic presentation of non-neoplastic endocervical cells. *Am J Clin Pathol* 1996;105:403–410.
8. Novotny DB, Maygarden SJ, Johnson DE, et al. Tubal metaplasia—a frequent potential pitfall in the cytologic diagnosis of endocervical glandular dysplasia on cervical smears. *Acta Cytol* 1992;36:1–10.
9. De Peralta-Venturino MN, Purslow MJ, Kini SR. Endometrial cells of the "lower uterine segment" (LUS) in cervical smears obtained by endocervical brushings: a source of potential diagnostic pitfall. *Diagn Cytopathol* 1995;12:263–271.
10. Johnson JE, Rahemtulla A. Endocervical glandular neoplasia and its mimics in Thin-Prep Pap tests: a descriptive study. *Acta Cytol* 1999;43:369–375.
11. Guidos BJ, Selvaggi SM. Detection of endometrial adenocarcinoma with the Thin-Prep Pap test. *Diagn Cytopathol* 2000;23:260–265.
12. Mody DR, Ramzy I. Glandular neoplasia or the uterus and adnexa. In: Ramzy I, ed. *Clinical Cytopathology and Aspiration Biopsy*, 2nd Ed. New York: McGraw-Hill, 2001:97–117.
13. Selvaggi SM. Cytologic features of squamous cell carcinoma in situ involving endocervical glands in endocervical brush specimens. *Acta Cytol* 1994;38:687–692.
14. Selvaggi SM. Cytologic features of high-grade squamous intraepithelial lesions involving endocervical glands on ThinPrep cytology. *Diagn Cytopathol* 2002;26(3): 181–185.
15. Wright TC Jr, Cox JT, Massad LS, et al. 2001 Consensus Guidelines for the management of women with cervical cytological abnormalities. *JAMA* 2002;287:2120–2129.
16. Ayer B, Pacey F, Greenberg M, Bousfield L. The cytologic diagnosis of adenocarcinoma in situ of the cervix uteri and related lesions: I. Adenocarcinoma in situ. *Acta Cytol* 1987;31:397–411.
17. Lee KR, Manna EA, Jones MA. Comparative cytologic features of adenocarcinoma in situ of the uterine cervix. *Acta Cytol* 1991;35:117–125.
18. Wilbur DC, Dubeshter B, Angel C, et al. Use of thin-layer preparations for gyneco-

logic smears with emphasis on the cytomorphology of high-grade intraepithelial lesions and carcinoma. *Diagn Cytopathol* 1996;14:201–211.
19. Young RH, Clement PB. Endocervical adenocarcinoma and its variants. Their morphology and differential diagnosis. *Histopathology* (Oxf) 2002;41:185–207.
20. Lee KA, Genest DR, Minter LJ, et al. Adenocarcinoma in situ in cervical smears with a small cell (endometrioid) pattern: distinction from cells directly sampled from the upper endocervical canal or lower segment of the endometrium. *Am J Clin Pathol* 1998;109:738–742.

Chapter 7

Other Malignant Neoplasms

Sana O. Tabbara and Jamie L. Covell

Background

Malignant neoplasms, other than squamous and adenocarcinoma, infrequently involve the uterine cervix but nevertheless may be seen in cervical cytologic preparations.[1-3] Most often these tumors are uncommon primaries arising in the uterine corpus or adnexae that appear in the cervical preparation, either as exfoliated cells or via direct sampling of tumors that involve the cervix or vagina by direct extension. Secondary or metastatic tumors to the uterine cervix occur rarely, owing to the nature of the lymphatic drainage and low vascularity of the uterine cervix.[2,4] In general, a definitive classification of the tumors described in this chapter may not be possible on cytologic preparations alone because of limited sampling and cytomorphologic overlap with other entities. However, familiarity with these entities is useful when an unusual tumor morphology is encountered.

Uncommon Primary Tumors of the Cervix and Uterine Corpus (Figs. 7.1–7.4)

Carcinomas

Spindle Squamous Cell Carcinoma (Fig. 7.1)

Spindle squamous cell carcinoma is a poorly differentiated variant of squamous cell carcinoma characterized by pleomorphic, spindled non-keratinizing cells with high mitotic activity.[5,6] The differential includes sarcoma and malignant melanoma with spindle cell features. Immunocytochemistry for cytokeratin may be helpful to demonstrate an epithelial origin.

Poorly Differentiated Squamous Carcinoma with Small Cells

Poorly differentiated squamous carcinoma with small cells morphologically resembles a high-grade squamous intraepithelial lesion and may also be confused with small cell undifferentiated carcinoma[5] (see below). The cells

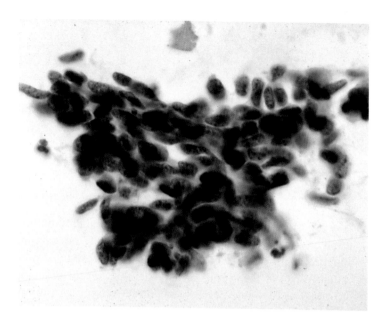

FIGURE 7.1. Spindle cell carcinoma (conventional preparation, CP). Spindle-shaped nonkeratinizing cells displaying variability in nuclear size, nuclear membrane irregularity, coarse granular chromatin, and conspicuous nucle-oli are arranged in a loosely cohesive cluster. The cytologic features are not specific and could be compatible with sarcoma, spindle cell carcinoma, or malignant melanoma.

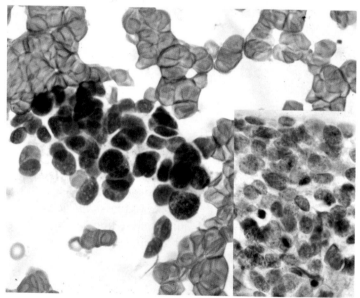

FIGURE 7.2. Small cell undifferentiated carcinoma (CP). Small to medium-sized cells with minimal cytoplasm, high nuclear/cytoplasmic ratio, hyper-chromatic nuclei, inconspicuous nucleoli, and prominent nuclear molding. The *lower right insert* shows the characteristic finely granular, stippled "neuroendocrine" chromatin pattern.

have more cytoplasm, greater cytoplasmic density, better definition of cell borders, and less crush artifact than small cell undifferentiated carcinoma.[3]

Small Cell Undifferentiated Carcinoma ("Oat Cell") (Fig. 7.2)

Small cell undifferentiated carcinoma (neuroendocrine carcinoma, grade III) comprises a small minority of all cervical carcinomas.[1,5] As in other body sites, this tumor is highly aggressive and is treated differently from other carcinomas. Therefore, it is important to recognize this tumor as distinct from poorly differentiated squamous carcinoma with small cells. Small cell undifferentiated carcinoma is composed of small, relatively uniform cells with scant cyanophilic cytoplasm. Characteristically, the cells are seen singly and in groups with nuclear molding, with "crush artifact" being a frequent finding. The nuclei are hyperchromatic with granular or stippled chromatin and inconspicuous nucleoli. Background tumor diathesis is common. Although the cytologic features of small cell undifferentiated carcinoma of the cervix are similar to those described in the lung and other body sites,[5,7–10] in the cervix these tumors are strongly associated with human papillomavirus (HPV) 18, a feature not identified in other primary sites.[10]

The differential includes poorly differentiated squamous carcinoma with small cells, poorly differentiated adenocarcinoma, low-grade endometrial stromal sarcoma, and lymphoma. The interpretation of small cell undifferentiated carcinoma should be reserved for tumors composed of small cells in which squamous or glandular differentiation is absent or minimal.[5] The presence of abnormal keratinized cells would favor an interpretation of poorly differentiated squamous cell carcinoma. If residual material from a liquid-based specimen is available, immunocytochemical staining for neuroendocrine markers [neuron-specific enolase (NSE), synaptophysin, chromogranin] may be useful to demonstrate neuroendocrine features.

Carcinoid Tumors

Carcinoid tumor (neuroendocrine carcinoma, grade I) is a rare primary tumor of the cervix. The small cells with high nuclear/cytoplasmic ratio resemble small cell undifferentiated carcinoma but lack nuclear molding.[4,11] Adenocarcinomas of the cervix may also demonstrate "carcinoid-like" features.[12]

Malignant Mixed Mesodermal Tumor (MMMT) or Carcinosarcoma (Figs. 7.3, 7.4)

Malignant mixed mesodermal tumor (MMMT) is an uncommon (<5% of malignant neoplasms of the uterine corpus) and highly aggressive car-

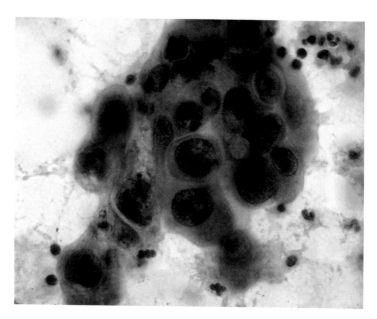

Figure 7.3. Malignant mixed mesodermal tumor (MMMT) (CP). Three-dimensional cluster of large epithelioid cells with round but pleomorphic nuclei, coarse granular chromatin, macronucleoli, and a moderate amount of cytoplasm.

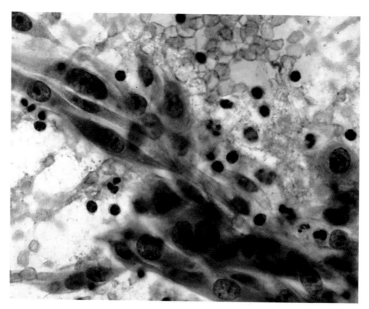

Figure 7.4. Malignant mixed mesodermal tumor (MMMT) (CP). Spindle cells with pleomorphic nuclei, coarse granular chromatin, macronucleoli, and a moderate amount of cytoplasm constitute the "sarcomatous" component of the same tumor depicted in Figure 7.3.

cinosarcoma that arises in the endometrium but may extend as a fungating mass into the cervical os. By definition, the tumor is biphasic, being composed of malignant epithelial and mesenchymal components. The malignant epithelial component morphologically most often resembles endometrioid adenocarcinoma. Mesenchymal (sarcomatous) elements are usually stromal, fibroblastic, or leiomyosarcomatous; occasional heterologous elements may include rhabdomyosarcoma, chondrosarcoma, or osteosarcoma. Recent clinicopathologic, immunohistochemical, and molecular genetic studies have provided strong evidence that MMMTs are best classified as variants of carcinoma.[13]

Exfoliated malignant cells from the endometrium or direct sampling of extension of an MMMT to the cervix/vagina may yield malignant cells in a cervical cytology sample. The presence of both malignant epithelial (Fig. 7.3) and sarcomatous components (Fig. 7.4) suggests the possibility of MMMT. However, degeneration or limited sampling of poorly differentiated malignant cells may lead to interpretive difficulties.[14–16] The differential diagnosis includes endometrial adenocarcinoma and pure sarcoma.

Sarcomas (Fig. 7.5)

Primary sarcomas of the female genital tract are rare; these can originate from the vagina, cervix, uterus, fallopian tubes, or ovaries but most commonly arise in the uterine corpus. Sarcomas may be pure or mixed with epithelial components and usually present with degenerated, sparse, or isolated tumor cells in the cervical sample.[1–3]

Pure sarcomas include leiomyosarcoma, endometrial stromal sarcoma, rhabdomyosarcoma, and fibrosarcoma. Most pure sarcomas present with undifferentiated, pleomorphic, multinucleated, and/or bizarre cells and cannot be further subtyped (Fig. 7.5). If present, characteristic cytologic features such as spindle or strap cells may suggest the specific type of sarcoma.[1–3,12]

Other Primary Tumors

Primary cervical germ cell tumors have been described, including choriocarcinoma, yolk sac tumor, and teratomas.[5] Leukemia/lymphoma and malignant melanoma rarely can be primary in the cervix.

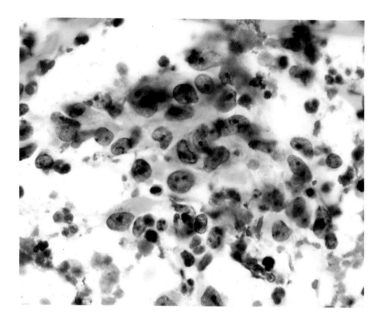

FIGURE 7.5. Sarcoma not otherwise specified (NOS) (CP). A loosely cohesive group of haphazardly arranged malignant cells with enlarged irregular nuclei and prominent nucleoli. Distinctive epithelial or mesenchymal differentiating features are not present.

Secondary or Metastatic Tumors

Extrauterine Carcinomas (Figs. 7.6–7.8)

Extrauterine carcinomas may spread to the cervix, or be present in a cervical cell sample, in one of three ways. Direct extension from a primary tumor in the pelvis, such as endometrium, bladder, and rectum, is the most common source of cervical involvement by secondary carcinoma.[5] Lymphatic and/or hematogenous metastases to the cervix are less frequent, with the most common primary sites being the gastrointestinal tract (Figs. 7.6–7.8), the ovary and the breast.[5] Last, exfoliated cells from an ovarian tumor (see Figs. 6.40–6.42) or from malignant ascites may pass through the fallopian tubes, endometrial cavity, and endocervical os, and end up in the cervical sample.

The majority of patients with metastatic tumors in a cervical sample have a prior history of a malignancy that leads to the correct interpretation.[3,5] Very rarely, cervical involvement is the first manifestation of disease. The metastasis may be recognized by its unique cytologic features or because the cells appear foreign to the preparation[17–21] (Table 7.1). The majority of metastatic tumors are characterized by a clean background

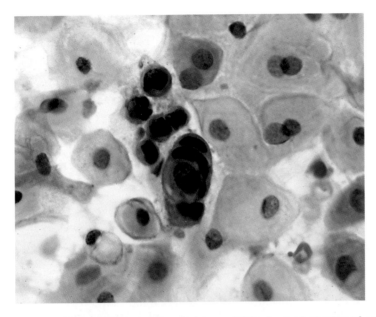

[∗] FIGURE **7.6.** Metastatic gastric carcinoma (CP). A small cluster of cells with malignant nuclear features displays the "cell in cell" arrangement often seen in gastric carcinoma. A cytoplasmic vacuole is present in one of the single cells. The background is free of tumor diathesis, a feature that favors metastatic rather than a primary origin of the tumor. [[∗] Bethesda Interobserver Reproducibility Project (BIRP) image (see xviii Introduction).]

TABLE **7.1.** Morphologic features of selected extrauterine carcinomas.

Primary site	Cytologic features
Breast	Signet ring cells Cell in cell arrangement
Stomach	Signet ring cells Cell in cell arrangement
Ovary and tube	Large cells Tight papillary clusters Psammoma bodies
Colon	Tall columnar cells with mucin
Kidney	Large cells Large round nuclei with macronucleoli Abundant delicate cytoplasm
Bladder	Similarity to squamous metaplastic epithelium Dense cytoplasm

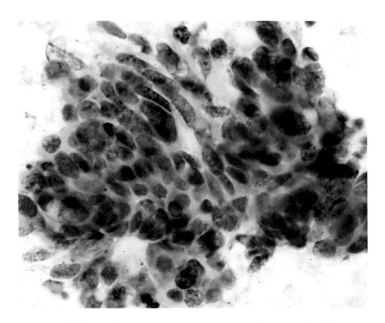

[∗] Figure 7.7. Metastatic colon carcinoma (CP). A group of tall columnar glandular cells demonstrates nuclear pleomorphism, hyperchromasia, cellular overlap, and loss of polarity within the cell group. These morphologic features would lead to a diagnosis of malignancy. The columnar cell shape, palisading cigar-shaped nuclei, and scattered goblet cells containing distended mucin-filled vacuoles seen in this image are distinctive morphologic features of colonic adenocarcinoma, as is "dirty necrosis" (not shown here).

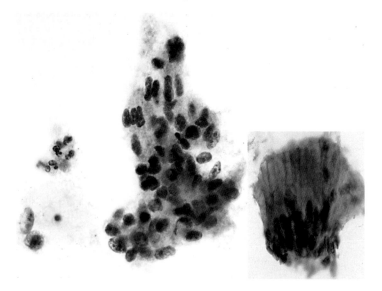

Figure 7.8. Metastatic colon cancer (liquid-based preparation, LBP). A cluster of malignant cells from metastatic colon carcinoma shows tall columnar cells with elongated nuclei at the upper edge and a glandular lumen in the center. Goblet cells are not identified in this group, and a mild degree of degeneration is noted. A fragment of normal colonic epithelium is shown in the *lower right insert* for comparison with the tumor cells in Figures 7.7 and 7.8.

or absence of tumor diathesis (see Fig. 6.41). However, when there is direct extension of tumor to the cervix/vagina, the associated tissue invasion and destruction can produce a tumor diathesis. Transitional cell carcinoma may involve the vagina by intraepithelial spread, and in such cases it may potentially be confused with squamous intraepithelial lesions and/or invasive squamous carcinoma.

Malignant Melanoma (Fig. 7.9)

Five percent to 10% of malignant melanoma in females arises in the genital tract on the vulva or vagina. Primary cervical melanoma is exceedingly rare, but metastatic melanoma is relatively more common.[5,22,23] The cytologic features are those common to melanoma from other sites. Cells are typically dissociated, and round, oval, or spindled in shape, with large nuclei and prominent nucleoli. Intranuclear pseudoinclusions and cytoplasmic melanin pigment may be seen. The differential diagnosis includes many poorly differentiated malignant neoplasms (primary or metastatic).

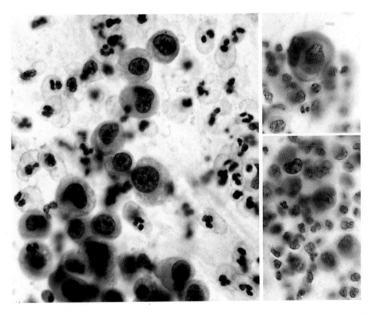

FIGURE 7.9. Malignant melanoma (CP). Dispersed and loosely cohesive large cells have a moderate amount of cytoplasm, round nuclei, irregular nuclear membranes, coarsely clumped, irregularly distributed chromatin, and prominent nucleoli. Cytoplasmic pigment consistent with melanin is a helpful finding but is not always present (*right upper* and *right lower* panels).

Immunocytochemical stains for S-100 protein, HMB45, and Melanin A may be useful.

Malignant Lymphoma

Malignant lymphoma may uncommonly involve the cervix in the context of disseminated disease or as a primary site.[24–26] The lymphoma cells are dispersed, and often show nuclear abnormalities including nuclear membrane irregularities and coarse uneven chromatin. An abnormal lymphoid population is generally more monotonous as compared to reactive chronic inflammatory processes; however, the specific morphology depends on the type of lymphoma. The differential diagnosis includes chronic/follicular (lymphocytic) cervicitis (see Figs. 2.39, 2.40) and small cell undifferentiated carcinoma. If a liquid-based specimen is available immunocytochemistry may be helpful in identifying a monoclonal lymphoid population.

References

1. Bonfiglio TA. Uncommon tumors of the cervix, vagina and uterine corpus. In: Bonfiglio TA, Erozan YS, eds. *Gynecologic Pathology*. Philadelphia: Lippincott-Raven, 1997:145–156.
2. Fu YS, Sherman ME. The uterine cervix. In: Silverberg SG, DeLellis RA, Frable WJ, eds. *Principles and Practice of Surgical Pathology and Cytopathology*, Vol. 3, 3rd Ed. New York: Churchill Livingstone, 1997:2343–2409.
3. DeMay RM. *The Pap Smear in the Art and Science of Cytopathology*, Vol. 6. Chicago: ASCP Press, 1996:136–137.
4. Kumar NB, Hart WR. Metastases to the uterine corpus from extragenital cancers. A clinicopathologic study of 63 cases. *Cancer* 1982;50:2163–2169.
5. Wright TC, Ferenczy A, Kurman RJ. Carcinoma and other tumors of the cervix. In: Kurman RJ (ed) *Blaustein's Pathology of the Female Genital Tract*, 5th Ed. New York: Springer-Verlag, 2002:325–381.
6. Steeper TA, Piscioli F, Rosai J. Squamous cell carcinoma with sarcoma-like stroma of the female genital tract. Clinicopathologic study of four cases. *Cancer* 1983;52:890–898.
7. Gersell DJ, Mazoujian G, Mutch DJ, et al. Small cell undifferentiated carcinoma of the cervix: a clinicopathologic, ultrastructural and immunohistochemical study of 15 cases. *Am J Surg Pathol* 1988;12:684–698.
8. Kamiya M, Uei Y, Higo Y, et al. Immunocytochemical diagnosis of small cell undifferentiated carcinoma of the cervix. *Acta Cytol* 1993;37:131–134.
9. Walker AN, Mills SE, Taylor PT. Cervical neuroendocrine carcinoma: a clinical and light microscopic study of 14 cases. *Int J Gynecol Pathol* 1988;7:64–74.
10. Stoler MH, Mills SE, Gersell DJ, et al. Small-cell neuroendocrine carcinoma of the cervix. A human papillomavirus type 18-associated cancer. *Am J Surg Pathol* 1991;15:28–32.

11. Miles PA, Herrera GA, Mena H, et al. Cytologic findings in primary malignant carcinoid tumor of the cervix, including immunohistochemistry and electron microscopy performed on cervical smears. *Acta Cytol* 1985;29:1003–1008.

12. Kurman RJ, Norris HJ, Wilkinson E. Tumors of the cervix, vagina, and vulva. In: *Atlas of Tumor Pathology*, Third Series, Fasc. 4. Washington, DC: Armed Forces Institute of Pathology.

13. Ronnett BM, Zaino RJ, Ellenson LH, et al. *Blaustein's Pathology of the Female Genital Tract*, 5th Ed. New York: Springer-Verlag, 2002:538–540.

14. Hidvegi DF, DeMay RM, Sorensen K. Uterine Mullerian adenosarcoma with psammoma bodies. Cytologic, histologic and ultrastructural studies of a case. *Acta Cytol* 1982;26:323–326.

15. Costa MJ, Tidd C, Willis D. Cervicovaginal cytology in carcinosarcoma (malignant mixed Mullerian [mesodermal] tumor) of the uterus. *Diagn Cytopathol* 1992;8:33–40.

16. Casey MB, Caudill JL, Salomao DR. Cervicovaginal (Papanicolaou) smear findings in patients with malignant mixed muellerian tumors. *Diagn Cytopathol* 2003;28:245–249.

17. Lemoine NR, Hall PA. Epithelial tumors metastatic to the uterine cervix: *Cancer* 1986;57:2002–2005.

18. Fiorella RM, Beckwith LG, Miller LK, et al. Metastatic signet ring carcinoma of the breast as a source of positive cervicovaginal cytology. A case report. *Acta Cytol* 1993;37:948–952.

19. McGill F, Adachi A, Karimi N, et al. Abnormal cervical cytology leading to the diagnosis of gastric cancer. *Gynecol Oncol* 1990;36:101–105.

20. Takashina T, Onto M, Kanda Y, et al. Cervical and endometrial cytology in ovarian cancer. *Acta Cytol* 1988;32:159–162.

21. Takashina T, Ito E, Kudo R. Cytologic diagnosis of primary tubal cancer. *Acta Cytol* 1985;29:367–372.

22. Holmquist ND, Torres J. Malignant melanoma of the cervix: a report of a case. *Acta Cytol* 1988;32:252–257.

23. Fleming H, Mein P. Primary malignant melanoma of the cervix. A case report. *Acta Cytol* 1994;38:65–69.

24. Whitaker D. The role of cytology in the detection of malignant lymphoma of the uterine cervix. *Acta Cytol* 1976;20:510–513.

25. Komaki R, Cox JD, Hansen PM, et al. Malignant lymphoma of the uterine cervix. *Cancer* (Phila) 1984;54:1699–1704.

26. Harris NL, Scully RE. Malignant lymphoma and granulocytic sarcoma of the uterus and vagina. *Cancer* 1984;53:2530–2545.

Chapter 8

Anal-Rectal Cytology

Teresa M. Darragh, George G. Birdsong, Ronald D. Luff,
and Diane D. Davey

Background

The use of anal-rectal cytology in the evaluation of human papillomavirus (HPV)-related lesions is a relatively new tool; its usefulness is still being investigated, demonstrated, and described.[1–3] It has been utilized in the evaluation of HPV-related disease of the anal canal, particularly in "high-risk" populations such as those who engage in anal intercourse and those with human immunodeficiency virus (HIV) disease.[4] The 1991 Bethesda system did not include other organ sites; however, there are parallels between cervical/vaginal and anal-rectal screening, and Bethesda System terminology has been used for reporting anal-rectal cytology.[1]

Sampling

The target of sampling includes the entire anal canal, the keratinized and nonkeratinized portions, and the anal transformation zone; the term "anal-rectal" was proposed to highlight the need to sample above the distal portion of the anal canal.

Cytologic samples are commonly collected without direct visualization of the anal canal, although some clinicians report using a small anoscope to introduce the collection device. No specific literature exists regarding the appropriate sampling device for anal cytology. Both Dacron fiber swabs and cytobrushes have been used for sampling.[5] The Dacron swab is recommended over a cotton swab because it releases its cellular harvest more readily and it has a plastic stick/handle that may be more appropriate for use with liquid-based sampling. Based on experience of one of the authors (T. Darragh), the Dacron swab is better tolerated by the patient than is the cytobrush.

Both conventional smears and liquid-based cytologic preparations are used. Some investigators have reported that liquid samples increase cell yield and also reduce compromising factors such as obscuring fecal material, air-drying, and mechanical artifacts.[6,7]

Adequacy (Figs. 8.1–8.4)

There is a paucity of literature regarding what constitutes an adequate anal-rectal sample, and the lower limits for adequate cellularity for anal cytology specimens have not been defined. Generally, the cellularity of an adequate anal sample approaches that of a cervical sample. As a guide, minimum adequacy cellularity is approximately 2000 to 3000 nucleated squamous cells for conventional smears, based on expert opinion. For liquid-based anal samples, this is approximately equivalent to 1 or 2 nucleated squamous cells per high-power field (hpf) for ThinPreps (with a diameter of 20 mm) and 3 to 6 nucleated squamous cells/hpf for SurePath (with a diameter of 13 mm), depending on the optical parameters of the microscope being used.

Typical cellular elements found on these preparations include anucleate squames, nucleated squamous cells, squamous metaplastic cells, and rectal columnar cells (Fig. 8.1). Presence of anal transformation zone components (rectal columnar cells and/or squamous metaplastic cells) should be reported as an indicator of sampling above the keratinized portion of the canal (Fig. 8.2).

Lack of cell preservation and contamination with bacteria/fecal material may compromise evaluation (Fig. 8.3). A sample composed predominantly of anucleate squames or mostly obscured by fecal material is unsatisfactory for evaluation (Fig. 8.4).

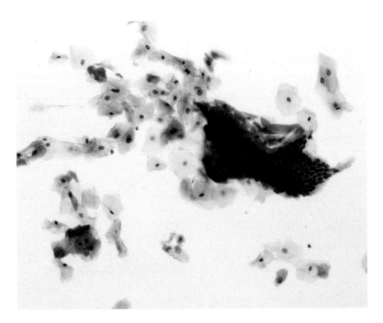

FIGURE 8.1. Satisfactory specimen, negative for intraepithelial lesion (NILM) (liquid-based preparation, LBP). Cellularity is adequate at low magnification. Both nucleated, well-preserved squamous cells and transformation zone component(s) are present.

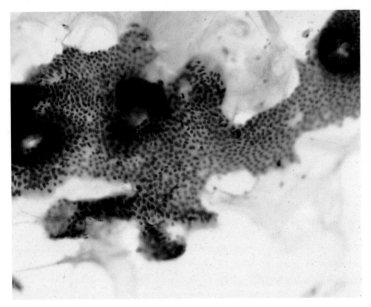

FIGURE 8.2. Normal rectal columnar cells (LBP). Just as with cervical cytology, the presence or absence of transformation zone component(s) should be indicated on an anal-rectal cytology report as a measure of specimen adequacy.

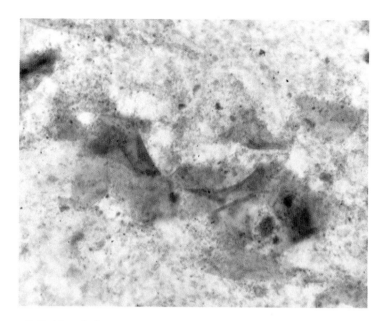

FIGURE 8.3. Unsatisfactory specimen (conventional preparation, CP). Particularly on conventional anal-rectal smears, bacteria and fecal material can obscure cellular detail.

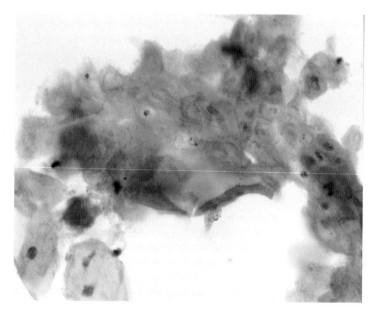

FIGURE 8.4. Unsatisfactory specimen (LBP). Predominantly anucleate squames.

Interpretation (Figs. 8.5–8.7)

Squamous intraepithelial lesions, squamous cell carcinoma, and infectious organisms can be identified in these specimens. Terminology, criteria, and guidelines for the evaluation of anal-rectal specimens should parallel those for cervical cytology.

The cytomorphologic criteria used for the evaluation of HPV-related anal lesions are analogous to those for cervical/vaginal cytology (Figs. 8.5–8.7). Degenerative cellular changes and squamous lesions with prominent orangeophilic cytoplasmic keratinization are common (Fig. 8.7) in anal-rectal specimens.

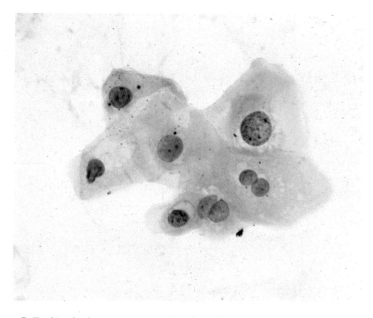

FIGURE 8.5. Atypical squamous cells of undetermined significance (ASC-US) (CP). Atypical "mature" squamous cells that do not meet the criteria required for an interpretation of low-grade squamous intraepithelial lesion (LSIL).

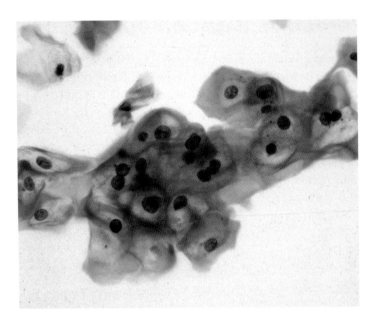

[∗] FIGURE 8.6. LSIL (LBP). Criteria for interpretation of SIL are the same as for cervical specimens (see Chapter 5). [[∗] Bethesda Interobserver Reproducibility Project (BIRP) image (see xviii Introduction).]

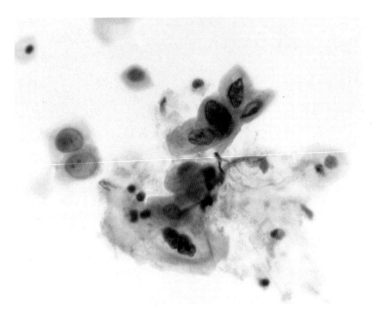

FIGURE 8.7. SIL (LBP). Both HSIL and LSIL are present in this figure. Note the cytoplasmic keratinization, a feature that is often more prominent in anal-rectal squamous lesions as compared to cervical squamous lesions.

Bethesda System 2001 Forum Group Moderators:

Diane D. Davey, M.D., George G. Birdsong, M.D., Henry Buck, M.D., Teresa M. Darragh, M.D., Paul Elgert, C.T. (ASCP), Michael Henry, M.D., Heather Mitchell, M.D., Suzanne Selvaggi, M.D.

References

1. Palefsky JM, Holly EA, Hogeboom CJ, et al. Anal cytology as a screening tool for anal squamous intraepithelial lesions. *J AIDS Hum Retrovir* 1997;14:415–422.
2. de Ruiter A, Carter P, Katz DR, et al. A comparison between cytology and histology to detect anal intraepithelial neoplasia. *Genitourin Med* 1994;70:22–25.
3. Scholefield JH, Johnson J, Hitchcock A, et al. Guidelines for anal cytology—to make cytological diagnosis and follow-up much more reliable. *Cytopathology* 1998;9:15–22.
4. Goldstone SE, Winkler B, Ufford BA, et al. High prevalence of anal squamous intraepithelial lesions and squamous cell carcinoma in men who have sex with men as seen in surgical practice. *Dis Colon Rectum* 2001;44:690–698.
5. Moscicki A, Hills NK, Shiboski S, et al. Risk factors for abnormal anal cytology in young heterosexual women. *Cancer Epidemiol Biomark Prev* 1999;8:173–178.
6. Darragh TM, Jay N, Tupkelewicz BA, et al. Comparison of conventional cytologic smears and ThinPrep preparations from the anal canal. *Acta Cytol* 1997;41:1167–1170.
7. Sherman ME, Friedman HB, Busseniers AE, et al. Cytologic diagnosis of anal intraepithelial neoplasia using smears and Cytyc Thin-Preps. *Mod Pathol* 1995;8:270–274.

Chapter 9

Ancillary Testing

Stephen S. Raab and Mark E. Sherman

Ancillary Testing

Briefly describe the assay performed and report the result clearly.

Background

The 2001 Bethesda System neither promotes nor discourages the use of ancillary testing in conjunction with cervical cytology; however, the workshop forum group members recognized the need to develop guidelines for reporting results of human papillomavirus (HPV) testing and other assays that might be implemented in the future.[1]

Following the Bethesda 2001 workshop, the American Society for Colposcopy and Cervical Pathology developed consensus management guidelines that suggest testing for oncogenic/high-risk types of HPV DNA to determine the management of "Atypical squamous cells of undetermined significance" (ASC-US) when the test can be performed using an already available sample.[2] This recommendation was based on results from ALTS[3] and other large studies[4,5] demonstrating that HPV testing was as sensitive as colposcopy in identifying women with an underlying cervical intraepithelial neoplasia (CIN) 2 or CIN 3, but reduced the number of referrals by nearly 50%, and data demonstrating the optimal cost-effectiveness of this approach.[6] The U.S. Food and Drug Administration has approved HPV ancillary testing for management of ASC, and in March 2003, approval was extended to include primary screening in conjunction with cytology among women 30 years of age and older. Using the specific example of HPV testing, recommendations for reporting ancillary tests in conjunction with cervical cytology are summarized below.

Description of Test Method and Result

The test method(s) should be briefly described (e.g., hybrid capture, polymerase chain reaction, in situ hybridization, etc.) and the results reported in a manner understandable to the clinician. For HPV testing, the specific types detected by the assay should be reported; testing should be restricted to oncogenic/high-risk types when possible.

Example 1: HPV hybrid capture assay–Positive for high-risk types of HPV. The HPV assay was performed using the [*Assay name*] [*Manufacturer name, City, State*]. The [*Assay name*] high-risk panel tests for HPV types: [*list HPV types*]. A positive test indicates the presence of one or more HPV types in the panel.

Example 2: In situ hybridization assay for HPV high-risk types is negative. The assay [*Assay name*][*Manufacturer name, City, State*] includes probes for HPV types [*list HPV types*].

Reporting of Molecular and Cytologic Results

It is preferable for cytology and ancillary test results to be reported concurrently to facilitate communication and record keeping. In addition, correlation of morpologic and ancillary test results can be a valuable tool for pathology education and ongoing quality assurance. However, not all clinical practice settings allow for integrated reporting of cytology and molecular results. If concurrent reporting is not feasible, then the report for each type of result should refer to the pending or previous report of the other test when possible.

Educational Notes and Suggestions

There are various models for reporting ancillary test data that may or may not incorporate management suggestions based on the combination of cytology and ancillary test results. Laboratories and clinicians should communicate their expectations regarding reporting of ancillary test results and inclusion of clinical management suggestions.[2] (See also Chapter 11, Educational Notes and Suggestions Appended to Cytology Reports.) Reference to published guidelines for using assay results may be useful in guiding patient management when appropriate.

Sample Reports

Examples of reporting are provided below.

Example 1: As a result only:
Interpretation: ASC-US with detection of oncogenic HPV DNA

Example 2: As a result associated with clinical management educational notes:
Interpretation: ASC-US with detection of high-risk oncogenic HPV DNA.
Note: The ASCCP consensus guidelines recommend that women with ASC-US who are HPV positive undergo colposcopic examination (*Wright TC et al. JAMA 2002;287:2120–2129*).

Example 3: As a result plus a probabilistic statement of an underlying CIN 2 or 3 *(the probabilistic reporting model)*:
Interpretation: ASC-US with detection of high-risk oncogenic HPV DNA.
Note: These findings are associated with a 15%–25% risk of underlying CIN 2 or higher grade lesion (*Cox JT et al. Am J Ob Gyn 2003; 188:1406–1412*).

Example 4: As a definitive interpretation that reflects both the cytomorphology and the HPV status *(the interpretive reporting model)*:
Interpretation: LSIL
Note: The preliminary cytologic findings are ASC-US. High-risk oncogenic HPV DNA has been detected. In combination, these results are most consistent with an interpretation of LSIL.

Bethesda System 2001 Workshop Forum Group Moderators:

Stephen S. Raab, M.D., Karen A. Allen, S.C.T. (ASCP), Christine Bergeron, M.D., Ph.D., Diane Harper, M.D., Walter Kinney, M.D., Alexander Meisels, M.D., Mark E. Sherman, M.D.

References

1. Solomon D, Davey D, Kurman R, et al. The 2001 Bethesda System terminology for reporting results of cervical cytology. *JAMA* 2002;287:2114–2119.
2. Wright TC Jr, Cox JT, Massad LS, Twiggs LB, Wilkinson EJ. 2001 management guidelines for the management of women with cervical cytologic abnormalities. *JAMA* 2002;287:2120–2129.
3. The ALTS Group. Results of a randomized trial on the management of cytology interpretations of atypical squamous cells of undetermined significance. *Am J Obstet Gynecol* 2003;188:1383–1392.
4. Manos MM, Kinney WK, Hurley LB, et al. Identifying women with cervical neoplasia: using human papillomavirus DNA testing for equivocal Papanicolaou results. *JAMA* 1999;281:1605–1610.
5. Bergeron C, Jeannel D, Poveda J, et al. Human papillomavirus testing in women with mild cytologic atypia. *Obstet Gynecol* 2000;95:821–827.
6. Kim JJ, Wright TC, Goldie SJ. Cost-effectiveness of alternative triage strategies for atypical squamous cells of undetermined significance. *JAMA* 2002;287:2382–2390.
7. Cox JT, Schiffman M, Solomon D, for the ALTS Group. Prospective follow-up suggests similar risk of subsequent cervical intraepithelial neoplasia grade 2 or 3 among women with cervical intraepithelial neoplasia grade 1 or negative colposcopy and directed biopsy. *Am J Obstet Gynecol* 2003;188:1406–1412.

Chapter 10

Computer-Assisted Interpretation of Cervical Cytology

Marianne U. Prey

Automated Review

If case is examined by automated device, specify device and result.

Background

Early attempts to objectively quantify microscopic images began with simple cell and nuclear measurements. In the 1960s, computers allowed for automation of this process and also permitted analysis of numerous other cellular features. Limitations of computing power hampered significant advancement in the field until the 1980s, when technological developments in computer hardware and artificial intelligence rekindled interest in automating cervical cytology screening.[1]

Different computer-automated slide scanning instruments are now used in various ways in the laboratory. Some instruments are designed to provide an independent assessment of the specimen; others are designed to interactively assist manual microscopy (e.g., by locating potentially abnormal areas for review). The following are general reporting recommendations for specimens evaluated by an automated device. These recommendations are not intended to be all-inclusive; additional relevant information may be included in the report, depending on the specific type of device employed.

Reporting the Results of Computer-Assisted Review

The preferred report format is to include a specific field designated for reporting appropriate information concerning the use of, and results from, the automated device. If this is not possible (e.g., because of laboratory information system constraints or local reporting convention), the automated screening information can be included as a comment or addendum. Some data resulting from automated review may not be intended for direct patient care but may be used for internal laboratory quality assurance; such data should not be included in the report.

The following information should be provided in the report:
1. Type of instrumentation used.
2. Whether or not the specimen was successfully *processed* by the device (regardless of the result).
3. Additional information depends upon whether there is manual screening/review of the specimen:
   a. *Automated Device Evaluation Without Manual Screening/Review*

   If the automated screening provides an interpretation of the specimen that replaces manual screening/review, then this result and any adequacy data derived from the computer assessment must be stated in the report.

   As with any automated laboratory instrument, the results generated by the instrument must be reviewed and verified by a laboratorian with appropriate training and authorization, even in the absence of manual screening/review. A record of who performed this data verification must be maintained as an internal laboratory record according to regulations issued pursuant to the Clinical Laboratory Improvement Amendments of 1988.[2] In general, the name of the individual performing such verification should *not* be included in the cervical cytology report, so as to avoid giving the false impression that the individual examined the specimen. However, if local laboratory policy requires inclusion of the name, the report should indicate that the individual did not examine the slide. The name of the medical director may be included as part of the laboratory identification per local custom and where required by state regulations.
   b. *Automated Screening Combined with Manual Screening/Review*

   If a specimen is manually screened or reviewed following automated screening, then results derived from the two methods must be compared. This comparison must be performed for both the adequacy evaluation and the interpretation, if such determinations are provided by the device. Any discrepancy must be reconciled before issuing the report. The name of anyone who examines a cervical cytology slide and renders an opinion for the final report should be documented in the report with the role of the person clearly stated.

Sample Reports

Example 1: Negative for intraepithelial lesion or malignancy, automated screen only:

Test Method:	Liquid-based cervical cytology
Source:	Cervix
Specimen Adequacy:	Satisfactory for evaluation, endocervical/ transformation zone component present
Interpretation:	Negative for intraepithelial lesion or malignancy
Automated Examination:	Processed successfully, manual screening not required [*Device name*] [*Manufacturer name, City, State*]

Example 2: Epithelial cell abnormality, with failure of automated screen:

Test Method:	Liquid-based cervical cytology
Source:	Cervix
Specimen Adequacy:	Satisfactory for evaluation, endocervical/ transformation zone component present
General Category:	Epithelial cell abnormality See Interpretation
Interpretation:	High-grade squamous intraepithelial lesion (HSIL) Fungal organisms morphologically consistent with *Candida* species
Automated Examination:	Processing failed, manual screening required [*Device name*] [*Manufacturer name, City, State*]
Educational Note:	Suggest further clinical investigation OR Suggest colposcopy and endocervical assessment (ASCCP guidelines: *JAMA* 2002:287:2120–2129)
Cytotechnologist:	CT (ASCP)
Pathologist:	Doctor, M.D.

Example 3: Epithelial cell abnormality, with successful automated screen
and manual screening:

Test Method:	Conventional Pap smear
Source:	Cervix
Specimen Adequacy:	Satisfactory for evaluation, endocervical/ transformation zone component absent
General Category:	Epithelial cell abnormality See Interpretation
Interpretation:	Atypical squamous cells of undetermined significance (ASC-US)
Automated Scanning:	Specimen processed successfully by automated locator device [*Device name*] [*Manufacturer name, City, State*]
Educational Note:	Suggest high-risk HPV testing as clinically indicated (ASCCP guidelines: *JAMA* 2002:287:2120–2129)
Cytotechnologist:	CT (ASCP)
Pathologist:	Doctor, M.D.

Bethesda System 2001 Workshop Forum Group Moderators:

Marianne U. Prey, M.D., Michael Facik, C.T. (ASCP), Albrecht Reith, M.D., Max Robinowitz, M.D., Mary Rubin, N.P., Ph.D., Sue Zaleski, S.C.T. (ASCP).

References

1. Koss L. *Diagnostic Cytology and Its Histopathologic Bases*. 4th Ed. Philadelphia: Lippincott, 1992.
2. Clinical laboratory improvement amendments of 1988: final rule. *Fed Reg* 1992;57: 493.1257 (Feb. 28).

Chapter 11

Educational Notes and Suggestions Appended to Cytology Reports

Dennis M. O'Connor

Educational Notes and Suggestions (optional)

Suggestions should be concise and consistent with clinical follow-up guidelines published by professional organizations (references to relevant publications may be included).

Background

Communication between laboratories and clinical providers is a key element of effective cervical cancer screening. Laboratories and clinicians share the responsibility of remaining current in their field and communicating significant changes in their respective disciplines to one another. Communication may take many forms: from informal conversations to formal grand rounds, journal articles, or presentations at meetings. One effective means of written communication is to append educational notes or suggestions to the cytology report. The method(s) of communication are left to the discretion of the laboratory and should be based on the individual practice setting and the content of the information to be conveyed.

Written comments regarding the validity and significance of a cytologic report are the responsibility of the pathologist and should be directed to the healthcare provider who requested the test. Generally, the laboratory should refrain from direct communication of results to the patient unless specifically requested by the provider and/or patient.

Educational Notes and Suggestions

Educational notes provide additional information regarding the significance or predictive value of the cytologic findings and may be based on references to the medical literature or the laboratory's experience. Appending educational notes or suggestions to the report is optional, but if used, the comments should be carefully worded, concise, clear, and evidence-based when possible.

For example, educational notes that highlight the limitations of cervical cytology as a screening test (previously referred to as "disclaimers") may be used on negative cytology reports. As another example, alerting clinicians to references containing consensus guidelines published by different professional organizations may be helpful because clinicians tend to follow opinions generated by their parent or allied medical societies.[1] Examples of guidelines pertinent to cervical cytology include those from the American Society for Colposcopy and Cervical Pathology (ASCCP),[2,3] the American College of Obstetricians and Gynecologists (ACOG),[4] the National Comprehensive Cancer Network (NCCN),[5] and the American Cancer Society (ACS).[6]

Appending suggestions (previously referred to as "recommendations") to the cytology report is also optional. If used, the format may vary depending on the preferences of the laboratory and its clinicians. The following examples highlight some circumstances in which suggestions could be helpful:

1. To improve the quality of a repeat specimen following receipt of an unsatisfactory specimen.
2. To identify patients with cytologic findings that may require further triage and management.
3. To indicate when further procedures would be helpful to clarify ambiguous morphologic findings.

Occasionally, clarification or amplification of a particularly complex report may require specific detailed suggestions. These points are best discussed directly with the provider before written comments are included in the report. If this is done, a sentence recording that the discussion took place is advisable (*"The significance of this report and possible management options were discussed with Dr. or Nurse _____ at _____ time on _____ date"*). If direct contact with the provider cannot be accomplished, general statements such as "Suggest follow-up as clinically indicated" or "Further diagnostic patient follow-up procedures are suggested as clinically indicated" should be used, because the pathologist may be unaware of other pertinent clinical information.

Sample Reports

Example 1:
   Specimen Adequacy:
   Satisfactory for evaluation, endocervical cells present.
   Interpretation:
   Negative for intraepithelial lesion or malignancy.
   Educational Note:
   Cervical cytology is a screening test primarily for squamous cancers
      and precursors and has associated false-negative and false-positive
      results. New technologies such as liquid-based preparations may de-
      crease but will not eliminate all false-negative results. Regular sam-
      pling and follow-up of unexplained clinical signs and symptoms are
      recommended to minimize false negative results.

Example 2:
   Interpretation:
   Specimen processed and examined, but unsatisfactory for evaluation of
      epithelial abnormality due to excessive air-drying artifact.
   Suggestion:
   Careful attention to rapid conventional slide fixation or the use of a
      liquid-based preparation is suggested to improve specimen quality.

Example 3:
   Interpretation:
   Atypical glandular cells, favor neoplastic.
   Educational Note:
   As a significant percentage of patients with this interpretation have un-
      derlying high-grade squamous or glandular intraepithelial abnor-
      malities, further diagnostic follow-up procedures are suggested as
      clinically indicated.
   (Optional addition of appropriate reference or references, for example,
      *Wright TC et al. 2001 Consensus guidelines for the management of
      cervical cytological abnormalities. JAMA 2002;287:2120–2129.)*

Bethesda System 2001 Workshop Forum Group Moderators

Dennis O'Connor, M.D., Marshall Austin, M.D., Ph.D., Lisa Flowers,
M.D., Blair Holladay, Ph.D., C.T. (ASCP), Dennis McCoy, J.D., Paul
Krieger, M.D., Gabriele Medley, M.D., Jack Nash, M.D., Mark Sidoti,
J.D.

References

1. Smith-McCune K, Mancuso V, Constant T, et al. Management of women with atypical Papanicolaou tests of undetermined significance by board-certified gynecologists: discrepancies with published guidelines. *Am J Obstet Gynecol* 2001;185:551–556.
2. Wright TC Jr, Cox JT, Massad LS, et al., for the 2001 ASCCP-sponsored Consensus Conference. 2001 Consensus guidelines for the management of cervical cytological abnormalities. *JAMA* 2002;287:2120–2129.
3. Davey DD, Austin RM, Birdsong G, et al. ASCCP patient management guidelines: Pap test specimen adequacy and quality indicators. J Lower Genital Tract Dis 2002;6: 195–199.
4. The ACOG Practice Bulletin "Cervical Cancer Screening" Number 45, August 2003.
5. Partridge EE. NCCN practice guidelines for cervical cancer screening. Version 2, 2003, pp 1–25. http://www.nccn.org/physician_gls/f_guidelines.html.
6. Saslow D, Runowicz CD, Solomon D, et al. American Cancer Society guideline for the early detection of cervical neoplasia and cancer. *CA Cancer J Clin* 2002;52: 342–362.

Index

Bold page numbers indicate Figures